T3-BUW-655

Contents

Preface

Environmentalists – ordinary people who wish to protect the earth for present and future generations – often voice concern about threats like acid rain, carbon dioxide buildup in the atmosphere, water contamination, and declining biodiversity. Recently, many have come to recognize another equally serious threat that we call "biological pollution."

We are a group of biological scientists who work in various ways to protect the plants that our species needs to survive and prosper. All of us are extremely concerned about biological pollution. In this publication, we describe our concerns and suggest ways in which control of biological pollution can be improved. We know that complete control is impossible, but we think the most serious consequences can be minimized.

WHAT IS BIOLOGICAL POLLUTION?

The problem we are talking about is the movement of living organisms, either accidentally or intentionally, from the places where they evolved to new environments where a lack of natural enemies permits their populations to explode. These organisms, sometimes called invasive exotic pests, threaten our crops, our forests, and perhaps our very existence.

Like chemical pollutants, biological pollutants are where they are primarily because of human activities. Chemical pollutants are produced directly by humans, and cause damage in direct proportion to the amount produced. The production of chemical pollutants can be reduced or prevented by legislation if the effects are sufficiently dire. That is not the case for biological pollutants. Once biological pollutants are imported, they grow, adapt, multiply and spread on their own unless people take direct, vigorous, and often costly actions to stop them.

Among scientists in many disciplines, concerns about exotic pests are growing. Few ecosystems are free of these pests, and several scientific conferences have been held recently to discuss the problems they cause. One such meeting was sponsored by the Forest Pathology Committee of the American Phytopathological Society (APS). Forest pathologists are especially concerned about imported pests because they have witnessed extremely severe damage by chestnut blight,

Dutch elm disease, and gypsy moths. Furthermore, new imported pests are creating new threats for forest trees.

At the meeting, scientists agreed that they need to reach a wider audience if the threat of biological pollution is to be minimized. This publication was mainly written by the speakers at the APS symposium. The intention here is to present a broad spectrum of concerns, in nontechnical language, to conservation-minded people.

Ultimately, deciding what to do about biological pollution in the United States is up to the American people. Our democratic system responds, albeit slowly, to mounting concerns of its citizens. Science can provide estimates of the hazards, and find ways to reduce the risks. But, at the very core of this issue are conflicting value judgements, weighing the many benefits of liberalized trade against the risk of environmental disaster. By raising public awareness of these risks, we hope to help legislators make informed decisions, and proceed with appropriate caution as they negotiate trade-enhancing treaties.

Our ultimate purpose is to ask for help. By ourselves, we who are paid to protect the world's forest ecosystems do not always succeed. Energized conservationists can increase our chances of success. As the various chapters in this publication show, success is very, very important.

The first three chapters of this book give some general background on biological pollution. They outline the enormity of the problem and the direct and indirect costs to date. The next three chapters provide specific examples of exotic weeds, diseases, nematodes, and insects and the damage they cause. The last four chapters examine the defenses that are in place to resist biotic invasion, and make suggestions on how these could be improved.

Acknowledgments

A number of friends and colleagues helped this manuscript evolve from scientific presentations into text that anyone interested in natural resources and agriculture might enjoy reading. Thanks are gratefully extended to Bob Biesterfeldt, Bruce Britton, Lynn Burgess, Carol Ferguson, Sharon Lumpkin, Sherry Trickel, and Arienna van Bruggen.

Controlling Biological Pollution

Kerry O. Britton
USDA Forest Service
Southern Research Station
Athens, GA*

Agriculture began as the cultivation of native plants, but it did not take people very long to discover that it is best to use the seeds from the best plants for the next planting. Over a long period, this kind of selection developed improved plants. There is little doubt that people traded improved plants and took them along when they moved. But movement then was in tens or, at most, hundreds of miles. Any problems caused by early introductions of plants to new areas are lost in the mists of time. In the last 2,500 years, people became increasingly mobile. They took their best plants with them, and brought the best plants back from new lands. These activities expanded the variety and productivity of food crops. Furthermore, as the human condition improved, the use of exotic plants as ornamentals became a passion, especially in Europe.

When Europeans explored and settled the Americas, they brought their improved plant varieties with them. Only wild ancestors of such plants as sunflowers, cranberries, blueberries, strawberries, pecans, hops, grapes, and some pasture grasses are thought to be native to the United States. On his second voyage, Columbus started the practice of bringing European plants to the New World. At the same time, the useful plants of the New World, including corn, potatoes, and tomatoes, were brought to Europe for cultivation.

In the nineteenth and early twentieth centuries, plant exploration was an official policy of the U.S. Government. For example, President John Quincy Adams ordered American diplomats to send home rare plant material. When the U.S. Department of Agriculture (USDA) was established in 1862, one of its objectives was to introduce new plants to America. An Office of Plant and Seed Introductions was created in the USDA in 1887. By 1923, the USDA had introduced some 50,000 plant species and varieties into the United States, many of them useful without question. Some, however, have proved to cause more harm than good. Introductions of new plants continue even now.

*Present address: USDA Forest Service, Forest Health Protection, Arlington, VA

Their pests often accompanied the new plants. At the time, introduction of pests didn't seem to be very important. We have learned from painful experience that they can be extremely damaging. Today, inspection of plant materials for agricultural pests is a routine practice. In the United States, thousands of pests are intercepted daily during border inspections. In many cases, finding these pests is fairly easy because closely related pest species attack the same crop species in many places. Sharing of information among countries helps in agricultural pest detection. But, forest pest problems are more difficult to predict, because forests are complex ecosystems; many hosts will interact with pests they have not encountered before.

Pest species slip through our inspection barriers, and there is the potential for big trouble when they do. Agricultural plants are particularly vulnerable because they are grown in large fields that contain only genetically similar members of the host species. If these host plants are susceptible, the pest population may grow rapidly and destroy much of the crop. Many farmers once rotated different crop species among their fields so that the same species was not grown in a field in successive years. Crop rotation reduced pest build-up. Now farmers often grow the same species on the same ground year after year, relying primarily on chemicals for pest control.

Fig. 1. Wisteria's seductive spring beauty conceals its destructive power. Imported as an ornamental, wisteria spreads by runners and by seed, smothering everything in its path.

Forest plantations produce wood fiber very rapidly, but they are highly vulnerable to pests because a single species covers a broad area. In the last 50 years, plantations of *Eucalyptus*, pine, willow, and poplar have been established half a world away from their native habitat. Freed from the pests of their native land, these trees often grow amazingly rapidly. But the damage can be equally amazing if the wrong pest shows up. Once again, the risk can be reduced by sharing of information about different species' pests. When they know what to look for, border inspectors often can intercept known pests.

AN IMPERFECT PROCESS

No matter how diligent the border inspectors are, the process that they are applying is imperfect. Travelers are often asked if they have any plant materials. If they are in a hurry and see little importance in the question, they may be less than truthful in their answers. Furthermore, it is next to impossible to compile a complete list of the pests that may cause damage in a country. Some pests, like nematodes, are very difficult to detect. They can be present in the soil and in plant products. Other pests cause little damage to their native hosts, but "jump" to a more susceptible species in a new land. The fungus that causes chestnut blight is a classic example.

Under these circumstances, some pests are going to slip through whatever barriers we erect. It therefore is also necessary to have effective means of eradicating exotic pests that make their way into our ecosystems. And the problem there is that the pests usually have a considerable head start before they are recognized as threats.

WEED SPECIES

A weed may be defined as a plant growing in a place where we do not want it, and an invasive weed species is one that is able to establish itself in many places where we do not want it. Some plants that were intentionally imported into the United States have become troublesome weeds. Kudzu and *Melaleuca* are two examples. Kudzu, a native of China, was widely planted in the United States to control erosion on bare soils. It did its job and then some, escaping to areas where we can get rid of it only with great difficulty. *Melaleuca* was brought to south Florida as an ornamental at the turn of the century. Its ability to seed in after the frequent fires in the region is prodigious, and it is occupying an average of about 50 new acres per day. It grows in extremely dense stands that change soil characteristics and drainage

Fig. 2. Yellow star thistle threatens to take over semi-arid rangeland and pastures in the western United States. It arrived as a contaminant of alfalfa seed, and spreads via crop seeds, hay and vehicle movement, and birds. It is poisonous to horses, and cattle dislike its long sharp spines.

patterns, destroying wildlife habitat and forage in and around the Everglades National Park.

Introduced bunchgrass and molasses grass now comprise 80% of the plant recover in parts of the Hawaii Volcanoes National Park. They burn easily and recover rapidly after fire. Frequent fires in the park increase the abundance of these grasses, causing spiraling destruction of native flora.

Wild hogs escaped from hunting enclosures in the Great Smoky Mountain National Park in 1912. Their descendants are destroying leaf litter habitat for smaller animals, increasing erosion, and consuming potentially threatened salamanders and snails in the park.

The opossum shrimp is devouring zooplankton in the Flathead River-lake ecosystem of Glacier National Park, causing declines in populations of fish and fish predators, including eagles, otters, coyotes, and bears (7).

E. O. Wilson, renowned Harvard socio-biologist, has said: "On a global basis . . . the two great destroyers of biodiversity are, first, habitat destruction and, second, invasion by exotic species. Extinction by habitat destruction is like death in an automobile accident: easy to see and assess. Extinction by invasion of exotic species is like death by disease: gradual, insidious, requiring scientific methods to diagnose" (8).

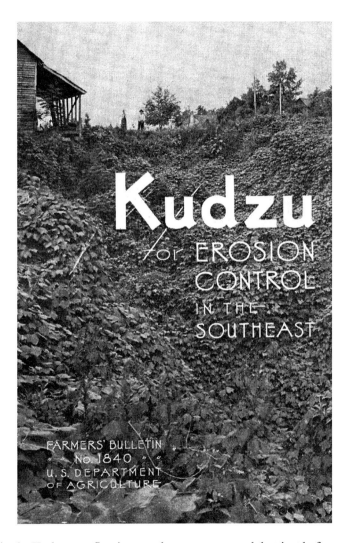

Fig. 3. Kudzu was first imported as an ornamental, but just before World War II farmers were paid $8 per acre to plant kudzu for erosion control.

MORE AND BETTER

How can we protect ourselves against biological pollution? Certainly, border inspections are part of the solution. Prompt eradications when unwanted species slip through are another part. But we are already doing these things. There seems little doubt that we must do what we are already doing, but that we must do it better.

We think that environmentalists can have a larger role. We need their understanding and their support. They can help us to spread the word that biological pollution is very serious. Research can provide the knowledge to deal rationally with the individual threats posed by biological pollutants. But research is expensive and its funding must be broadly supported. Environmentalists can provide that support. And they may be able to help us directly in our efforts to eradicate imported pests.

The introduction of exotic species has enormously increased the productivity of agriculture in the United States. Exotic ornamentals have given us more beautiful and varied landscapes. But it may be time to have a new look at importing plant species. Maybe we should be considering much stronger limitations on these introductions. Certainly, they should be attempted only when the effects are predictable. Again, environmentalists can assist in the political process in which laws are changed, and in publicizing the value of native plants in landscaping.

If you care about the problems that we describe in this publication, you can help. In fact, we cannot succeed without your help.

Literature Cited

1.   Exotic Pests of Eastern Forests. 1998. Britton, K.O., ed. Keen Impressions, Asheville, NC. 198 pp.

2.   Carefoot, G.L., and Sprott, E.R. 1967. "Famine on the Wind". Rand McNally and Co., New York. 229 pp.

3.   Horsfall, J.G. 1983. Impact of introduced pests of man. Pp. 13 *in:* Exotic Plant Pests and North American Agriculture. Wilson, C.L. and Graham, C.L., eds. Academic Press. New York. 522 pp.

4. Large, E.C. 1940. The Advance of the Fungi. Henry Holt and Co. New York. 488 pp.

5. McKnight, B.N. 1993. Biological Pollution: The Control and Impact of Invasive Exotic Species. Indiana Acad. Sci. Indianapolis. 261 pp.

6. Schoulties, C.L., Seymour, C.P., and Miller, J.W. 1983. Where are the exotic disease threats? *in*: Exotic Plant Pests and North American Agriculture. Wilson, C.L., and Graham, C.L., eds. Academic Press. New York. 522 pp.

7. U.S. Congress, Office of Technology Assessment, "Harmful Non-Indigenous Species in the United States". OTA-F-565. Washington, DC: U.S. Government Printing Office, September 1993.

8. Wilson, E.O. 1997. Foreword. *in*: Strangers in Paradise. Simberloff, D., Schmitz, D.C., and Brown, T.C., eds. Island Press, Washington, D.C. 467 pp.

An Ecological Explosion in Slow Motion

Randy G. Westbrooks
United States Department of Interior
US Geological Survey
Whiteville, NC

Peter White
University of North Carolina
Chapel Hill, NC

In recent years, many scientists have become aware of the threats that invasive species pose for natural and managed ecosystems (7,9, 17,22,23,27,28,29). It is clear that the extent of worldwide transport of species by humans is unprecedented (14). According to British ecologist Charles Elton (3), human-induced biological invasions represent ". . . one of the great historical convulsions of the world's flora and fauna. . . . There is no doubt that we are living in a period of the world's history when mingling of thousands of kinds of organisms from different parts of the world is setting up terrific dislocations in nature."

In this article, we describe how the changes that are taking place could affect numbers of species existing in the world. The estimates are necessarily rough, but there is little doubt that massive reductions in numbers of species are likely to occur.

BIOGEOGRAPHICAL REALMS THREATENED BY HUMAN ACTIVITIES

In the mid-1800s, British naturalist Alfred Russel Wallace (25) delineated six major biogeographical realms on the earth's land surfaces. These realms, which roughly correspond with the continents, have distinct assemblages of plants and animals. These species evolved in ecological and genetic isolation over the past 180 million years since the breakup of the supercontinent Pangaea (2,3). Evolution proceeded differently in each realm, developing unique arrays of species.

While colonizing virtually all of the habitable land on earth, humans have radically altered the biogeographical realms. As hunter-gatherers, numbers of humans were held in check by diseases, an unstable food supply, and other environmental factors that still limit the populations of other species today. With the development of agriculture, science, and medicine, however, humans became the earth's ultimate invader species (24). With colonization came the relocation of numerous domesticated plants and animals, as well as a long list of unintended hitchhikers. In comparison with the slow evolutionary changes of the past, the effects of human activities are explosive.

The earliest recorded plant collection expedition was undertaken in Asia Minor by the Sumerians in about 2500 B.C. (15). It was colonists from Western Europe, however, that created massive impacts over large areas by distributing plant and animal species. It is sobering to think that Native American tribes inhabited Florida for at least 10,000 years before European colonists introduced cultivated corn, beans, peppers and other vegetables from the nearby Caribbean region (30). As the human population has continued to grow over the last few centuries, the effects of its movement of species have escalated into one of the earth's most significant ecological events.

IMPACT OF EXOTIC INVADERS

When humans move species out of their native habitats, co-evolved parasites and predators normally are left behind. Free of enemies, the immigrant species is able to outcompete and displace native species. It may do so by altering water tables, changing natural fire regimes, suppressing reproduction of native species, and harboring diseases that kill the native species. It also may modify nutrient cycling regimes, change soil fertility, increase erosion, and provide habitat for other undesirable species (5).

At least 4,500 introduced species of plants and animals have established populations in the United States (12). About 675, or 15%, cause severe harm. Between 1906 and 1991, 79 harmful species caused $97 billion in documented losses in the United States (12). Since 1980, at least 205 additional invading species have been detected in the United States, and at least 59 of them are expected to cause serious economic and environmental harm (12). In the Western United States, introduced weeds are spreading at a rate of about 4,600 acres per day in what can be described as an ecological explosion in slow motion (1). Table 1 gives eight examples of invading species that are causing serious damage.

Table 1. Origin, affected region, and effect of some harmful non-indigenous species

Anopheles gambiae	Africa	Brazil, 1929	Severe outbreaks of malaria in the 1930s	Elton (3)
European starling (*Sturnus vulgaris*)	Europe	United States, 1891	Widespread pest, by the mid-1950s	Elton (3)
Kudzu [*Pueraria montana* var. *lobata* (Willd.) Ohwi]	Japan, China	Southeastern United States, 1876	Vine with dense canopy that shades out other plants	Shurtleff and Aoyagi (19)
Sea lamprey (*Petromyzon marinus*)	North Atlantic Ocean and northeastern rivers	U.S. Great Lakes, 1829	Major parasite of soft-bodied fish	Elton (3)
Witchweed [*Striga asiatica* (L.) O. Kuntze]	African grasslands	North Carolina and South Carolina, USA, 1956	Severe threat to grain production in the United States	Sand et al. (16)
Zebra Mussel (*Dreissenia polymorpha* Pallas)	Central European rivers	U.S. Great Lakes, mid-1980s, via ballast water from ships	Cover river and lake bottoms and municipal and industrial water inlets	Stolzenburg (21)
Water Hyacinth (*Eichhornia crossipes*) [Mart. Solms.]	American	Waterways in the southern United States, 1884, New Orleans Exposition	Floating acquatic weed that chokes waterways	Schmitz et al. (18)
Tropical Soda Apple (*Solanum viarum* Dunal)	Argentina, Brazil	Southeastern United States, 1988	Pasture production	Mullahey and Colvin (10)

What if, in effect, all species had equal access to all portions of the world? Arguably we are creating something close to that situation by carrying plants and animals wherever we choose. In that case, biologically, the earth would become a single continent again. After interspecies competition had run its course, how many species would remain?

No one can answer that question with much precision, but a rough estimate can be made. Preston (13) related the number of bird species to the amount of land area available. The more land area, the more species are likely to be present. He estimated that if all the land on earth were in a single mass the size of all continents combined, about 2,300 bird species would be present. With relatively isolated continents, we currently have over 8,600 species of birds on earth. He concluded that the isolated continents have permitted about four times as many species of birds as would be present if all the continents were joined.

Using the same approach, we calculated that a single continent would support about 2,000 species of mammals. Currently, the separated continents support about 4,200 species. We conclude that a complete breakdown of biogeographic barriers could lead to extinction of more than half of the continental mammals and an even larger fraction of the island species (Fig. 1).

These calculations are highly theoretical. They represent what could happen rather than what will happen. They are, however, as soundly based as calculations of species losses caused by habitat fragmentation from land-use changes (22,23).

Fossil records provide supporting evidence for the notion that numbers of species decline when continental barriers break down. About three million years ago, North and South America became connected by the Isthmus of the Panama land bridge. That narrow connection permitted massive exchanges of plants and animals between the two continents. While some South American species (notably the opossum) crossed the Isthmus into North America, many more North American mammals spread into South America. Associated with the influx of North American mammals was a significant increase in the extinction rate for South America's native mammals.

Mammals on Earth

Actual and Projected Species

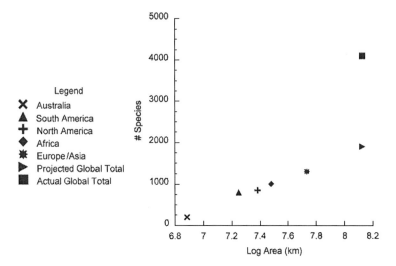

Fig. 1. Species of mammals on earth, by continent, with actual and projected global totals.

Conclusions

At present, human activities are reconnecting the continents in a biological sense. Natural historian Alfred Crosby (2) has said: "The breakup of Pangaea was a matter of geology and the stately tempo of continental drift. Our current reconstitution of Pangaea by means of ships and aircraft is a matter of human culture and the careening, accelerating, breakneck rate of technology."

There is no doubt that human transport of species among continents will lead to reductions in biodiversity. It should be emphasized, however, that the reductions we have calculated represent what could happen rather than what will happen. Human beings have the ability to observe negative trends and do something about them.

What we must do is to prevent further degradation of natural ecosystems and maintain the general integrity of Wallace's biogeo-graphical realms. These goals are a necessary challenge for government regulators, commercial importers, and the traveling public in all nations.

Literature Cited

1.   Asher, J. 1995. Proliferation of invasive alien plants on western federal lands: An explosion in slow motion. Pp. 5-9 *in:* Proceedings of the Alien Plant Invasions: Increasing Deterioration of Rangeland Ecosystem Health Symposium. January 17, 1995. Phoenix, AZ. Bureau of Range Mgt. Portland, Oregon. 63 pp.

2.   Crosby, A. 1986. Ecological Imperialism. The biological expansion of Europe, 900-1900. Cambridge University Press. NY. 368 pp.

3.   Elton, C. 1958. The ecology of invasions by animals and plants. Methuen and Co., Ltd. London. 181 pp.

4.   Ewel, J. 1986. Invasibility: Lessons from South Florida. Pp. 214-230 *in:* Mooney, H., and Drake, J., eds. Ecology of Biological Invasions of North America and Hawaii. Springer-Verlag, NY. 361 pp.

5.   Jensen, D., and Vosick, D. 1994. Introduction. Pp. 1-6 *in:* Schmitz, D., and Brown, T., eds. An assessment of invasive non-indigenous species in Florida's public lands. FL. Dep. Env. Prot. Bur. Aqu. Plt. Mgt. Tallahassee. Tech. Rept. TSS-94-100. 303 pp.

6.   Marshall, L., Webb, S., Sepkowski, J., and Raup, D. 1982. Mammalian evolution and the great American interchange. Science 215:1351-1357.

7.   McKnight, B., ed. 1993. Biological pollution: The control and impact of invasive exotic species. Proc. Symp. on Biological Pollution. Ind. Acad. Sci. Indianapolis. October 25-26, 1991. 261 pp.

8.   Milne, L., and Milne, M. 1977. Ecology out of joint. New environments and why they happen. Charles Scribner & Sons. NY. 304 pp.

9. Mooney, H., and Drake, J., eds. 1986. Ecology of biological invasions of North America and Hawaii. Springer-Verlag, NY. 361 pp.

10. Mullahey, J., and Colvin, D. 1993. Tropical soda apple: A new noxious weed in Florida. Univ. Florida Fact Sheet WRS-7. Gainesville. 3 pp.

11. Old, R. 1992. Noxious weeds invading nation at an alarming rate. Vegetation Manager VI:4.

12. OTA. 1993. Harmful non-indigenous species in the United States. U.S. Congress, Office of Technology Assessment. OTA-F-565. Washington, D.C. U.S. Government Printing Office. 391 pp.

13. Preston, F. 1960. Time and space and the variation of species. J. Ecol. 41:611-627.

14. Primack, R. 1993. Essentials of conservation biology. Sinauer Assoc., Inc. Sunderland, MA. 564 pp.

15. Ryerson, K. 1967. The history of plant exploration and introduction in the United States Department of Agriculture. Pp. 1-19 *in:* Proc. Intl. Symp. on Plant Introduction. Held in Tegucigalpa, Honduras on November 30–December 2, 1966. Escuela Agricola Panamericana, Honduras.

16. Sand, P., Eplee, R., and Westbrooks, R., eds. 1990. Witchweed research and control in the United States. Monograph. Weed Science Society of America. Champaign, IL. 154 pp.

17. Schmitz, D., and Brown, T., eds. 1994. An assessment of invasive non-indigenous species in Florida's public lands. FL. Dep. Env. Prot. Bur. Aqu. Plt. Mgt. Tallahassee. Tech. Rept. TSS-94-100. 303 pp.

18. Schmitz, D., Nelson, B., Nall, L., and Schardt, J. 1991. Exotic aquatic plants in Florida: A historical perspective and review of the present aquatic plant regulation program. Pp. 303-326 *in:* Center, T., Doren, R., Hofstetter, R., Myers, R.,

and Whiteaker, L., eds. Proc. Symp. On Exotic Pest Plants. Univ. Miami, Rosentiel School of Marine and Atmospheric Science. November 2-4, 1988. U.S. Dept. of Interior/Natl. Park Serv., Washington, DC.

19. Shurtleff, W., and Aoyagi, A. 1977. The Book of Kudzu. Autumn Press, Brookline, MA.

20. Silver, T. 1990. A New Face on the Countryside. Indians, Colonists, and Slaves in South Atlantic Forests, 1500-1800. Cambridge University Press. NY. 204 pp.

21. Stolzenburg, W. 1992. The mussels' message. Nature Conservancy 42:16-23.

22. Vitousek, P., D'Antonio, C., Loope, L., and Westbrooks, R. 1996. Biological invasions as global environmental change. American Scientist (September-October Issue) 84:468-478.

23. Vitousek, P., D'Antonio, C., Loope, L., Rejmanek, M., and Westbrooks, R. 1997. Introduced species: A significant component of human-caused global change. New Zealand Journal of Ecology 21(1):1-16.

24. Wagner, W. 1993. Problems with biotic invasives: A biologist's viewpoint. Pp. 1-8 *in:* McKnight, B., ed. Biological pollution: The control and impact of invasive exotic species. Proc. Symp. on Biological Pollution. Ind. Acad. Sci. Indianapolis. October 25-26, 1991. 261 pp.

25. Wallace, A. 1876. The Geographical Distribution of Animals. London, 2 vols.

26. Westbrooks, R. 1981. Introduction of foreign noxious plants into the United States. Weeds Today 14:16-17.

27. Westbrooks, R. 1991. Plant protection issues. I. A commentary on new weeds in the United States. Weed Technology 5:232-237.

28. Westbrooks, R. 1992. Regulatory exclusion of Federal Noxious Weeds from the United States by USDA APHIS.

Pp. 110-113 *in:* Proc. 1st Intl. Weed Control Congress, Feb.
17-21, 1992. Monash University, Melbourne, VIC, AUS.

29.   Westbrooks, R. 1993. Exclusion and eradication of
        foreign weeds from the United States by USDA APHIS.
        Pp. 225-241 *in:* Biological pollution: The control and impact
        of invasive exotic species. McKnight, B., ed. Proc. Symp.
        on Biological Pollution. Ind. Acad. Sci. Indianapolis.
        October 25-26, 1991. 261 pp.

30.   Winters, H. 1967. The mechanics of plant introduction.
        Pp. 49-53 *in:* Proc. Intl. Symp. On Plant Introduction. Held
        in Tegucigalpa, Honduras on November 30–December 2,
        1966. Escuela Agricola Panamericana, Honduras.

Exotic Pests:
Past, Present, and Future

Phyllis N. Windle
Windle Research Services
College Park, MD

In 1990, I directed the U.S. Congressional Office of Technical Assessment (OTA) study of exotic pests. After several years of data gathering and analysis involving more than 100 expert contractors, advisory panelists, and peer reviewers, we concluded that the threat to our nation's resources is serious and continuing. We had hoped to find that protection had improved since chestnut blight and Dutch elm disease reached our shores, but we were disappointed. Exotic pests continue to enter the United States at a worrisome rate. Between 1980 and 1993, 205 exotic species were reported as introduced or detected (19). Since no single organization is responsible for tracking newly introduced species, and our OTA attempt was limited, that total of 205 species probably was an underestimate. At least 59 of the new species are known or expected to be harmful. Clearly, therefore, much remains to be done if we are to protect our country's resources from exotic pests.

HOW THEY GOT HERE

The cumulative number of exotic species in the United States has risen swiftly over the past 200 years (Fig. 1). At least 4,500 species of foreign origin have established free-living populations here (19). And that total is certainly a minimum. As locales are studied in detail, the number climbs. For example, one study found 212 exotic species in the San Francisco Bay area alone (3).

Despite assertions that the pace of introductions is increasing dramatically, OTA (19) found no clear evidence of an increase over the past 50 years. Instead, the number of new introductions appears to fluctuate, probably in response to social, political, and technological factors (Fig. 2).

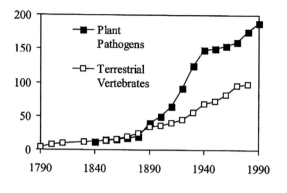

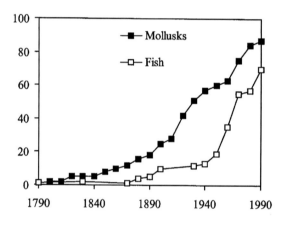

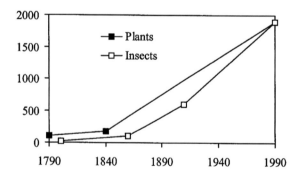

Fig. 1. Documented accumulating introductions of exotic species.
Reprinted from U.S. Congress (19).

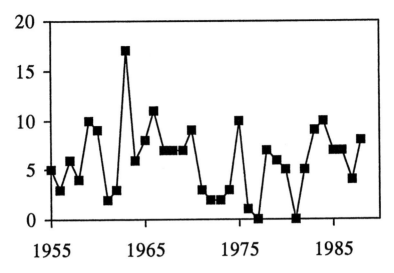

Fig. 2. Insect and other invertebrate species established in California. Reprinted from U.S. Congress (19).

New state and federal plant quarantine laws slowed the introduction of insect pests and plant pathogens after 1912. In the 1800s, the switch from dry to wet ship ballast decreased weed introductions but increased introductions of aquatic organisms. In Norfolk and Baltimore harbors, for example, ships dump more than 10 million tons of ballast water each year, and one study found living organisms in ballast water in 90 percent of incoming vessels (2).

ENTRY PATHWAYS

Exotic species enter the United States by four major pathways:

1. as hitchhikers on imported plants, animals, and various products;

2. as intentionally imported species not intended for distribution;

3. as intentionally imported species intended for distribution, but with unknown damaging properties; and

4. as illegal trade material.

The first pathway is the most common, and agricultural commodities are the most frequent carriers. Between October 1987 and mid-July 1990, 81% of the noxious weeds found by federal inspectors were associated with agricultural commodities (24). And all 23 of the insect species established in California since 1980 arrived with imported commodities – 20 on plants, 2 on fruit, and 1 on wood. OTA's list of 205 species entering since 1980 included 108 with known routes of entry. Of these, 83 hitchhiked with seeds, plants, fruit, wood, packing material, or ballast water.

Fig. 3. Mile-a-minute vine is also called devil's tail tearthumb, because barbs lurk on the stem and underside of leaves. It arrived on ship's ballast in Oregon, and contaminated rhododendrons shipped to Pennsylvania.

The second pathway has provided entry for non-native crop species, ornamentals, livestock, pets, and aquaculture species. Inspection and quarantine procedures are designed primarily to guard against the unintentional import of insects, pathogens, and other pests associated with the plants or animals that are intentionally imported. At the outset, most of these intentional imports are contained in some manner, but some of them break from containment and cause very serious problems. For example, about 300 plant species are invading wildlands in the United States and Canada; about half of these were imported as garden plants (11). In Florida, 90% of the most invasive plants with known modes of entry were intentionally imported. About half were intended to be ornamentals (8). Likewise, escapes from the aquarium and aquaculture industries have spread exotic aquatic weeds, fish, fish pathogens, and fish parasites into the wild. Some experts predict that species imported for aquaculture are virtually guaranteed to escape to the wild (Walter R. Courtenay, Jr., personal communication) (15,16).

Invasive species that followed the third entry pathway include kudzu, peacock cichlids, and wild hogs. They were species that appeared to be beneficial but later proved to be very harmful.

The final mode of entry – illegal introduction – may be attributable largely to ignorance of the risks on the part of importers. In Florida, for example, there appears to be a thriving illegal trade in reptiles and amphibians (4). Plants and animals are often carried in the baggage of returning or visiting passengers. The U.S. Department of Agriculture typically intercepts more than 1 million pieces of luggage containing unauthorized living materials each year (19).

Strategies for preventing entry vary with the entry pathway. Enormous diligence is required of border inspectors to prevent hitchhiking. Often, however, it may be possible to prevent hitchhikers by treating commodities prior to export. Such treatments would have to be agreed upon by trading nations. The second and third pathways could be narrowed by strengthening rules for intentional importation of species. Much of the illegal traffic might be halted through educational programs to show people how harmful imported species can be.

Until the early 1900s, private individuals usually made decisions about whether to introduce exotic species. There was little or no government oversight. Even when government officials were involved, decisions were informal and often lenient. Over the years, more formal decision-making methods have emerged, including risk analysis, legally mandated environmental impact assessment, and cost/benefit analysis.

PRESENT IMPACTS

Highly damaging exotic species are present throughout the country. Costly impacts are felt in agriculture, forestry, human health, and the protection of natural areas. The zebra mussel, for example, causes massive private and public losses. Public utilities must remove them from clogged water intakes; landowners must clear them from irrigation channels, and fish and wildlife agencies must try to control them to maintain the health of aquatic systems.

About 40% of exotic species cause some harm, and 15% cause severe economic or environmental damage. Intentional introductions, which usually receive some screening prior to entry, usually cause less harm than unintentional introductions. Still, about 12% of intentional introductions prove harmful.

The OTA found that just 79 introduced species cost $97 billion between 1906 and 1991 (19). That total is conservative because no economic data are available for many minor pests, and because intangible, non-market impacts are ignored. Environmental impacts often are difficult to evaluate. Recent research suggests that a highly virulent exotic disease in Australia (10) and introduced predators in California's Central Valley (6) are at least partially responsible for the worldwide decline in frog populations. How can one estimate the economic importance of this event?

In the long run, species that alter ecosystem properties may have greater impact than those that affect a single native species. For example, *Melaleuca* has converted grasslands in the Florida Everglades into single-species forests. In the West, cheat grass invasions have altered the magnitude and frequency of wildfire, causing further changes in the species present on grassland.

DECIDING WHICH SPECIES ARE SAFE

Restrictions on intentional importing of species are becoming more stringent. Many introductions that once might have been encouraged are no longer allowed. Import procedures vary considerably, depending on what types of resources may be at risk. Risks to nonagricultural areas often are ignored, and new imports are generally presumed to be safe unless proven otherwise. Most regulatory agencies use some type of list that characterizes some species or groups as "dirty" and therefore prohibited; others are "clean" and therefore permitted. For more than 20 years, biologists have asserted that these short blacklists of species allow too much exposure to harm (14,19). Thus far, however, efforts to change these features of laws have failed.

Unfortunately, there are still no reliable predictors of a given species' potential as an invader. Each decision about a new species' import and release, therefore is hampered by uncertainty. Three interrelated problems must be addressed: (1) determining acceptable risks; (2) setting thresholds for risk and other factors, above which more formal and costly decision-making approaches are required; and (3) identifying and making tradeoffs when deciding in the face of uncertainty.

EMERGING HUMAN AND WILDLIFE DISEASES

At present, regulations to prevent the transport of wildlife and human diseases are weak. Perhaps that is why the exotic fish parasite that causes whirling disease has spread widely in the United States. In parts of Montana, it has destroyed as much as 90% of the rainbow trout population (9). In 1992, Newcastle Disease was found in double-crested cormorants; it killed at least 5,000 wild birds in the United States and Canada (12,23). The National Wildlife Health Research Center has documented a number of waterfowl disease outbreaks that have killed 25,000 to 100,000 birds (7). Some were caused by microorganisms imported into the United States; others by pathogens spreading beyond their usual U.S. distribution.

At least 1,465 Americans fell ill in 1996 from a *Cyclospora* parasite first identified in New Guinea in 1979 (1). One recent invader, the Asian tiger mosquito, is likely to cause substantial health problems in the United States. This species reached Texas in 1985 probably via imports of used tires from Japan. By the summer of 1994, it had reached at least 23 other States. It carries several serious tropical diseases, including dengue fever, yellow fever, and several kinds of encephalitis. There were few and less effective vectors for these diseases in the United States until the introduction of the tiger mosquito. For instance, eastern equine encephalitis, which kills up to 75% of its human victims, so far has been successfully limited to wildlife and domestic animals in the United States, because its usual vector bites only animals. In contrast, the Asian tiger mosquito feeds on both animals and humans and can move the disease back and forth freely.

TRADE AND EXOTIC PESTS

Experts have recognized a large number of potential pests not yet in the United States. The Federal Noxious Weed Act currently lists 89 such species, but a technical advisory group to APHIS recommended in 1983 that 750 others be added. Potential economic losses from just 15 high-impact exotics could be as high as $134 billion (19). Clearly, losses of that magnitude cannot be ignored.

At the same time, the value of global trade continues to increase. The value of world trade in agricultural products rose 5-fold from about $40 billion in 1970 to more than $200 billion in the early 1990s (22). Over the same period, agricultural inspections at U.S. borders increased 20-fold, and APHIS anticipates another doubling by the year 2000.

So far, international environmental treaties have not been a major force limiting the global spread of harmful exotic species. International environmental laws, like the Convention on Biological Diversity and the International Plant Protection Convention, have been adopted, but their provisions are weaker than national prerogatives already in effect in the United States.

International trade agreements provide another potential but largely unrealized means of regulating movement of harmful species. Exotic species simply have not been a major consideration in the General Agreement on Tariffs and Trade (GATT), its new forum for dispute resolution, the World Trade Organization (WTO), or the North American Free Trade Agreement (NAFTA). Nevertheless, trade agreements would seem to be an ideal medium for protocols for preventing the spread of harmful species.

Conclusion

There is no shortage of exotic pests in the United States now, and many more can be expected if barriers against them are not improved. The potential monetary losses from pests that have not yet arrived are astronomical. Clearly, the time has come for effective action.

Acknowledgments

Much of the work discussed here was done while I directed OTA's study of exotic species. The assessment team included analytical staff Elizabeth Chornesky, Peter T. Jenkins, Steven Fondriest, and

Kathleen E. Bannon. We also had the administrative help of Nathaniel Lewis, Nellie Hammond, and Carolyn Swann, plus work from numerous expert contractors, advisory panelists, and reviewers.

Literature Cited

1.   Boodman, S.G. 1997. Forbidding fruit: How safe is our produce? The Washington Post. July 8, pp. 10-13.

2.   Chesapeake Bay Commission. 1995. The Introduction of Nonindigenous Species to the Chesapeake Bay Via Ballast Water. Annapolis, MD.

3.   Cohen, A.N., and Carlton, J.T. 1995. Biological Study. Nonindigenous aquatic species in a United States Estuary: a case study of the bological invasions of the San Francisco Bay and Delta. U.S. Fish and Wildlife Service, Arlington, VA.

4.   Cox, J.A., Quinn, T.G., and Boyter, H.H. 1997. The Role of the Florida Game and Fresh Water Fish Commission in the Management of Nonindigenous Species. Pages 297-316 *in*: Strangers in Paradise: Impact and Management of Florida's Nonindigenous Species. Simberloff, D., Schmitz, D.C., and Brown, T.C., eds. Island Press, Washington, D.C.

5.   Culotta, E. 1994. The weeds that swallowed the West. Science 265:1178-1179.

6.   Fisher, R.N., and Shaffer, H.B. 1996. The Decline of Amphibians in California's Great Central Valley. Cons. Biol. 10(5): 1387-1397.

7.   Friend, M. 1992. Environmental influences on major waterfowl diseases. Trans. 57th N. Amer. Wildlife & Natl. Res. Conf., Pages 517-525.

8.   Gordon, D., and Thomas, K.P. 1997. Introduction of invasive nonindigenous plants into Florida: history, screening, and regulatory approaches. Pages 21-37 *in*:

Strangers in Paradise: Impact and Management of Florida's Nonindigenous Species. Simberloff, D., Schmitz, D.C., and Brown, T.C., eds. Island Press, Washington, D.C.

9. Jenkins, P.T. 1996. Free trade and exotic species introductions. Cons. Biol. 10(1):300-302.

10. Laurence, W.F., McDonald, K.R., and Speare, R. 1996. Epidemic disease and the catastrophic decline of Australian rain forest frogs. Cons. Biol. 10(2):406-413.

11. Marinelli, J. 1996. Introduction: redefining the weed. Pages 4-6 *in*: Invasive Plants. Randall, J.R., and Marinelli, J., eds. Handbook #149. Brooklyn Botanic Garden, Brooklyn, NY.

12. Nettles, V.E. 1996. Re-emerging and emerging infectious diseases: economic and other impacts on wildlife. U.S. House of Representatives Committee on Agriculture and the American Society for Microbiology.

13. Nickum, D. 1996. Whirling disease in the United States: overview and guidance for research and management. Trout Unlimited, Arlington, VA.

14. Ruesink, J.L., Parker, I.M., Groom, M.J., and Kareiva, P.M. 1995. Reducing the risks of nonindigenous species introductions. Biosci. 45(7):465-477.

15. Shelton, W.L., and Smitherman, R.O. 1984. Exotic fishes in warm water aquaculture. Pages 262-302 *in*: Distribution, Biology, and Management of Exotic Fishes. Courtenay, W.R., and Stauffer, J.R., eds. The Johns Hopkins University Press, Baltimore, MD.

16. U.S. Aquatic Nuisance Species Task Force. 1992. Proposed aquatic nuisance species program. Draft report to Congress.

17. U.S. Centers for Disease Control and Prevention. 1995. Reptile associated salmonellosis – selected states, 1994-1995. Morbidity & Mortality Weekly Rept. 44(17):347-350.

18.   U.S. Congress, Office of Technology Assessment. 1992. Trade and Environment: Conflicts and Opportunities. OTA-BP-ITE-94. U.S. Government Printing Office, Washington, D.C.

19.   U.S. Congress, Office of Technology Assessment. 1993. Harmful Nonindigenous Species in the United States. OTA-F-565. U.S. Government Printing Office, Washington, D.C.

20.   U.S. Congress, Office of Technology Assessment. 1995. Agriculture, Trade, and Environment: Achieving Complementary Policies. OTA-ENV-617. U.S. Government Printing Office, Washington, D.C.

21.   U.S. Congress, General Accounting Office. 1996. Commercial Trucking, Safety and Infrastructure Issues Under the North American Free Trade Agreement. GAO/RCED-96-01. Washington, D.C.

22.   U.S. Department of Agriculture. 1994. Desk Reference Guide to U.S. Agricultural Trade. Agriculture Handbook No. 683, Foreign Agricultural Service, Washington, D.C.

23.   U.S. Department of the Interior. 1993. Newcastle disease virus found in double-crested cormorants in the United States. U.S. Fish and Wildlife Service Research Info. Bull. No. 13.

24.   Westbrooks, R.G. 1991. Federal noxious weed inspection guide: noxious weed inspection system. U.S. Animal and Plant Health Inspection Service, Washington, D.C.

Exotic Weeds:
Expensive and Out of Control

Randy G. Westbrooks
United States Department of Interior
US Geological Survey
Whiteville, NC

Robert E. Eplee
USDA APHIS
Oxford Plant Protection Center
Oxford, NC

Particularly for people who do not see farms very often, weeds do not seem like a modern, high-tech problem. But they are a serious problem if you grow crops for a living. Weeds always have been and probably always will be the first barrier for agriculture. If you want to grow a crop, the first thing you have to do is clear away the vegetation that would otherwise compete with the crop. And any farmer can tell you that weeds were not created equal. Some species are a lot more trouble than others. Thus, the last thing a farmer needs is a new species of weed that grows faster than the ones that are present now.

Introduced weeds that are causing big problems in the United States include witchweed in the Carolinas, kudzu throughout the Southeast, mile-a-minute vine in the Northeast, leafy spurge in the Midwest, and miconia in Hawaii. Introduced plants like them receive little attention until they create major problems (6). By the time a problem is identified, and funding is obtained for control, eradication often is impractical. By then, the invasive plant has become a permanent, expanding, and detrimental component of the ecosystem.

ECONOMIC IMPACT OF WEEDS

In agriculture, weeds cause two kinds of costs: (1) monies spent for control, and (2) reductions in crop production caused by competition with the crop species. During the 1950s, such losses in the United States were estimated to be $5.1 billion per year (17). By 1994, such costs had risen to at least $20 billion per year. Some $15 billion were associated with crops, pasture, hay, and range, and $5 billion with golf

Fig. 1. Leafy spurge produces toxins that inhibit the growth of native vegetation in prairies and pastures, and is difficult to kill with herbicides. A complex of natural enemies has been imported from Europe for biological control.

courses, turf and ornamentals, highway rights-of-way, industrial sites, forests, and other sites. The values of losses were not available for most non-crop sites, but control costs were estimated for them. Since introduced species account for about 65% of the weed flora in the United States, their total impact on the U.S. economy is at least $13 billion per year (Fig. 1).

Losses to individual species, such as leafy spurge, are quite impressive. In 1991, researchers reported that direct annual losses to that one weed were $2.2 million in Montana, $78 million in North Dakota, $1.4 million in South Dakota, and $200,000 in Wyoming. Nearly 6% of all the untilled land in North Dakota is infested with leafy spurge (9), which cows cannot eat. Total direct and indirect impacts of leafy spurge on grazing land, wildlife, water conservation and the State's economy in North Dakota totaled in excess of $87 million annually (8).

It should also be mentioned that weed losses in the agricultural sector would probably increase about 5-fold without the use of herbicides (2,3). In 1995, U.S. farmers spent about $5 billion on herbicides.

ROLE OF THE FEDERAL GOVERNMENT
IN WEED MANAGEMENT

Various federal agencies assist with weed management problems in the United States. Within the U.S. Department of Agriculture, the Animal and Plant Health Inspection Service (APHIS) tries to prevent foreign weeds from entering the United States, and, failing in that, from becoming established on private land here. APHIS cooperates with state and local agencies, as well as private landowners, in eradicating newly introduced weed species on private land. Research on agricultural weeds is conducted by the USDA's Agricultural Research Service. On federal land weed control and some research to support that effort is conducted by the federal land management agencies: the USDA Forest Service, the U.S. Fish and Wildlife Service, the National Park Service, the Bureau of Land Management, the Bureau of Reclamation, the U.S. Geological Survey, the Bureau of Indian Affairs, the Department of Defense, and the Department of Energy.

With so many agencies involved in weed control, close cooperation is needed for operations on federal land. Similar degrees of cooperation are needed on private land. Here, state and local governments as well as landowners must be involved.

The APHIS strategy for control of imported noxious weeds includes:

Prevention – encouraging the production of pest-free commodities in foreign countries.

Preclearance – inspection and certification of certain commodities at the port of export, prior to shipment to the United States.

Exclusion – inspections and treatments at the port of entry to detect and exterminate prohibited pests.

Detection – conducting surveys and communicating with scientists and state officials to rapidly detect incipient infestations of prohibited foreign species.

Containment – establishment and enforcement of rules and programs to prevent the spread of prohibited species from infested areas to non-infested areas.

Eradication – Total elimination of infestations of prohibited species by appropriate means.

Biological control – control of certain pests with biological agents if eradication fails.

FEDERAL NOXIOUS WEEDS LIST

In 1976, 26 foreign weeds were placed on a list for exclusion from the United States. APHIS inspectors at ports of entry were given this list and procedures for inspecting commodities to detect noxious weeds. At that time, raw wool, soil-contaminated equipment, aquatic plant shipments, and seed shipments were recognized as likely carriers of weed seeds (21,23).

In the late 1980s, APHIS adopted a system to help its inspectors to detect weed contamination in high-risk commodities. This system provides officers with information on potential associations of target weeds and commodities that originate in places where target weed species can be expected to grow. Useful references are kept at inspectors' work stations (7,21,23,24).

In 1996, the list of noxious weeds was expanded to include 89 species. Many argue that this list should be expanded even further.

EARLY DETECTION AND RAPID RESPONSE

Despite the best efforts of inspectors at ports of entry, invaders pass through and find soil in which to grow. The next step in protection, therefore, is to attempt to detect the invader before it can spread. If it is detected early, an infestation can be contained and eradicated at a relatively low cost.

In the past, weed specialists were not systematically informed about the presence of new species. They often learned about invaders by word of mouth or through notes published in botanical journals. Communication could be improved by creating a plant detection network in every state. The network would include plant collectors, herbarium curators, botanists, farmers, county agents, and land managers as well as state and federal weed control specialists.

To speed control efforts, state weed teams are being established in Pennsylvania, North Dakota, Wyoming, Utah, New Mexico, and Delaware. The teams are meant to coordinate defensive action and to quickly bring needed resources to bear. We think all states should consider establishing such teams. We also think a federal interagency rapid response weed team would be useful. When an infestation is too firmly established to permit eradication, ways are needed to prevent its spread. Ways are also needed to control large infestations. Research is probably the key to success in preventing spread and controlling large infestations.

Conclusion

Over the past 40 years, APHIS and the agencies that preceded it have achieved numerous successes. Infestations of imported weed species have been eradicated from many locations in the United States. Nevertheless, weed invaders are slipping through. We have suggested some ways in which the performance of our defenses would be improved. First, the list of potentially damaging invaders could be expanded considerably. Second, early detection of new infestations could be improved through closer communication among land managers, amateur and professional botanists, and weed specialists. Third, each state could establish a weed team to coordinate actions when an infestation is detected. A federal interagency rapid response team also would be helpful. Finally, research will be needed to develop effective ways to control large infestations and to prevent their spread.

Literature Cited

1.   American Crop Protection Association. 1996. American Crop Protection Association Industry Profile for 1995. American Crop Protection Association, Washington, D.C.

2.   Bridges, D., ed. 1992. Crop losses due to weeds in the United States. Weed Sci. Soc. Am., Champaign, Il. 403 pp.

3.   Bridges, D. 1994. Impact of weeds on human endeavors. Weed Technology 8:392-395.

4. Chandler, M. 1985. Economics of weed control in crops. Pp. 9-20 in: ACS Symposium Series, No. 268. The Chemistry of Allelopathy Biochemical Interactions Among Plants. Thompson, A., ed. American Chemical Society.

5. Elton, C. 1958. The ecology of invasions by plants and animals. Methuen and Co., Ltd. London, England. 181 pp.

6. Eplee, R., and Westbrooks, R. 1990. Federal Noxious Weed Initiatives For The Future. Proc. Weed Sci. Soc. NC. pp 76-78.

7. Eplee, R., and Westbrooks, R. 1991. Recent advances in exclusion and eradication of Federal Noxious Weeds. WSSA Abstracts 31:31.

8. Goold, C. 1994. The high cost of weeds. Pages 5-6 *in:* Noxious weeds: Changing the face of southwestern Colorado. San Juan National Forest Association, Durango, Colorado.

9. Leistritz, F., Bangsund, D., and Leitch, J. 1995. Economic impact of leafy spurge on grazing and wildland in the northern Great Plains. Pages 15-21 *in*: Alien plant invasions: Increasing deterioration of rangeland ecosystem health. BLM/OR/WA/PT-95/048+1792. Proceedings of a symposium by the Range Management Society, Phoenix, Arizona.

10. McKnight, W., ed. 1993. Biological Pollution: The control and impact of invasive exotic species. Proc. Symp. Biological Pollution. Ind. Acad. Sci. Oct. 25-26, 1991.

11. Montgomery, F. 1964. Weeds of Canada and the northern United States. Ryerson Press, Toronto, Canada. 226 pp.

12. Mooney, H., and Drake, J., eds. 1986. Ecology of biological invasions of North America and Hawaii. Springer-Verlag, New York. 321 pp.

13. Ross, M., and Lembi, C. 1983. Applied Weed Science. Burgess Publishing Company. 340 pp.

14   Schmitz, D. 1990. The invasion of exotic aquatic and wetland plants in Florida: History and efforts to prevent new introductions. Aquatics 12:6-13, 24.

15.   Schmitz, D. 1994. The ecological impacts of non-indigenous plants in Florida, *in:* Schmitz, D., and Brown, T., eds. An assessment of invasive non-indigenous species in Florida's public lands. Fl. Dep. Env. Prot. Tech. Rpt. TSS-94-100, 10-28.

16   Shaw, W. 1979. National Research Program 20280, weed control technology for protecting crops, grazing lands, aquatic sites, and noncropland. Weeds Today 10(4):4.

17.   USDA. 1965. A survey of extent and cost of weed control and specific weed problems. U.S. Department of Agriculture, Agricultural Research Service, Washington, D.C. Report # 23-1. 78 pp.

18.   Vitousek, P., D'Antonio, C., Loope, L., and Westbrooks, R. 1996. Biological invasions as global environmental change. American Scientist 84(5):468-478.

19.   Wagner, W.H., Jr. 1993. Problems with biotic invasives: biologist's viewpoint, in: McKnight, B.N., ed. Biological Pollution: The control and impact of invasive exotic species. Proc. Symp. Biological Pollution. Ind. Acad. Sci. Oct. 25-26, 1991, pp. 1-8.

20.   Westbrooks, R. 1981. Introduction of foreign noxious plants into the United States. Weeds Today 14:16-17.

21.   Westbrooks, R. 1989. Regulatory exclusion of Federal Noxious Weeds from the United States. Ph.D. Dissertation. Department of Botany, N.C. State University, Raleigh, NC. 335 pp.

22.   Westbrooks, R. 1991. Plant Protection Issues. I. A commentary on new weeds in the United States. Weed Technology 5:232-237.

23.   Westbrooks, R., and Eplee, R. 1991. USDA APHIS
       Noxious Weed Inspection System. 1991 Update. WSSA
       Abstracts 31:29.

24.   Westbrooks, R. 1993. Exclusion and eradication of foreign
       weeds from the United States by USDA APHIS, *in:*
       McKnight, B.N., ed. Biological Pollution: The control and
       impact of invasive exotic species. Proc. Symp. Biological
       Pollution. Ind. Acad. Sci. Oct. 25-26, 1991, pp. 225-241.

25.   Westbrooks, R., and Eplee, R. 1996. Strategies for
       preventing the world movement of invasive plants. Pp. 29-
       35 *in*: Proceedings of the United Nations–Norway
       Conference on Alien Species. Trondheim, Norway. July 1-
       5, 1996. Norwegian Ministry of the Environment. Oslo,
       Norway.

26.   Westman, W. 1990. Park management of exotic plant
       species: Problems and Issues. Cons. Bio. 4:251-260.

27   Zamora, D., Thill, D., and Eplee, R. 1989. An eradication
       plan for plant invasions. Weed Technology 3:2-12.

Plant Disease on the Move!

Kerry O. Britton
USDA Forest Service
Southern Research Station
Athens, GA*

Frank Tainter
Clemson University
Clemson, SC

Judith Brown
University of Arizona
Tucson, AZ

Plant diseases are caused by the same types of pathogens as human and animal diseases: bacteria, viruses, fungi, nematodes, and mycoplasma-like organisms. Unlike human pathogens, they evolved ways to spread that did not rely on host movements. Most succeeded by producing massive amounts of reproductive spores or eggs that spread passively in wind, rain, and moving water. A few have found vectors, mostly sucking insects like aphids, whiteflies, and leaf-hoppers, which inject their propagules into new hosts, much as the *Anopheles* mosquito spreads malaria.

Recently, plant pathogens have accomplished long-distance transport with the ultimate vector, *Homo sapiens*. Migration rates for plants have increased from something like 10 miles per century, for natural range expansion and shrinkage, to 500 miles per hour on the average jet airliner. And, if these plants are infected, their pathogens arrive with all the reproductive capacity they developed to spread within their previously immobile host population. This chapter describes a few classic examples of how economically and ecologically costly the results can be.

*Present address: USDA Forest Service, Forest Health Protection, Arlington, VA

LATE BLIGHT OF POTATO

A classic example of devastation by an exotic plant disease is the great Potato Famine of Ireland. This disaster was caused by a fungus imported on potatoes from South America, where the plant is native. The famine that followed starved a million Irish people between 1845 and 1848, induced two million more to emigrate, and forever altered the population structure of the United States. One reason the loss of potatoes was so devastating was that the Irish depended on potatoes for about 80% of their caloric intake. This lack of diversity in their diet left the Irish extremely vulnerable. Potato blight and the ensuing starvation have been blamed for the downfall of British Prime Minister Robert Peel, and the repeal of the English 'corn laws', which restricted imports and protected the incomes of English landowners. The relaxation of these trade policies increased the strength of Great Britain as a world power. Potato blight spread, though less intensely, throughout Europe. The resulting hunger and widespread unrest, particularly in the cities, provided "fertile ground for the radical seeds being sown by Marx and Engels", and added fuel to revolutions in France, Hungary, and Austria (6).

In 1882, a mixture of copper sulfate, lime, and water, which had been applied to grapes in France to deter small boys from pilfering, was found by accident to control most fungal plant diseases, including the potato blight. Thus, by the close of the nineteenth century, a cure for potato blight had been developed. But, in Germany during World War I, the copper needed for this fungicidal mixture was diverted to the manufacture of shell casings and electric wire. Then the cool wet summer of 1916 provided ideal conditions for the exotic fungus to infect the unprotected potato crop, and 700,000 Germans died of starvation. This famine accelerated the defeat of Germany on the western front (6).

Today, well-developed countries may seem less vulnerable to such epidemics, because most diets have become much more diverse. However, almost all of the crop plants we grow in the United States came from other countries (13), just as the potato was exotic to Ireland. A host of disease and insect pests lurk in the countries of origin of our crop species.

In general, host plants develop some resistance mechanisms that limit their destruction by co-evolved pests. So why are agricultural crops vulnerable to pests from their country of origin? One reason is that they lack genetic diversity.

Low biodiversity certainly contributed to the devastation of the 'Irish' potato. Potatoes are reproduced by tuber pieces, so cultivars are essentially clones. Since each neighboring plant in a potato field is genetically identical, a fungus to which they are susceptible can easily spread from one to another. Blight-resistant cultivars have been developed by returning to wild potatoes for resistance genes. Developing resistant cultivars is often economically feasible in agricultural crops. Since most crop species are annual plants, their relatively short reproductive cycle lends itself to rapid genetic improvement. Recent advances in molecular biology have further accelerated the use of genetic resistance to control exotic pests.

CHESTNUT BLIGHT

At the beginning of the 20th century, American chestnut trees were a major component in 25% of the hardwood forests of the eastern United States. This magnificent tree provided highly valued timber, and a nut crop for wildlife, domestic animals, and people.

In 1904, H.W. Merkel, chief forester and founder of the New York Zoological Society, discovered that a fungus disease was killing 438 chestnut trees in the Bronx parks. At first, it was thought that severe drought had made the trees more susceptible to a native, usually harmless fungus. But soon scientists noticed that Chinese and Japanese chestnuts were not affected very much by the disease. Such resistance usually indicates that the host species and the pathogen have co-evolved. The fungus probably hitch-hiked on Asian chestnut nursery stock, which had been imported for landscape plantings during the previous century. Massive sanitation programs were undertaken to eradicate the disease, but it spread inexorably (14).

In 1938, chestnut blight was reported on the European chestnut in Italy. The disease spread throughout Europe much as it had in the United States. Europe braced for disaster. But soon the Italians noticed that the cankers caused by the blight fungus (*Cryphonectria parasitica*) were healing over. The European strains of the fungus had developed a virus disease that weakened the fungus and permitted the trees to survive. European chestnut stands are alive and producing nuts today (15).

Much effort has been directed toward spreading this virus disease into the American strains of *C. parasitica*. For many years, only some of the American strains were susceptible, but in 1996, pathologists succeeded in getting virus particles into the spores, and spreading the

Fig. 1. Fred Hebard displays inoculated resistant chestnut tree at the American Chestnut Foundation Research Farm.

virus to some strains that were previously not susceptible. Thus, hope is renewed that the chestnut blight fungus can be weakened in America (7,8).

A second approach to chestnut blight has been to incorporate resistance genes from Asian chestnuts into American chestnut trees (Fig. 1). By back-crossing the hybrids to American chestnut three times, a tree with 15/16 American chestnut genes has been produced. Intercrosses of the progeny are currently being tested for continued resistance, and may be available commercially within just a few years. (See http://www.chestnut.acf.org for periodic updates).

WHITE PINE BLISTER RUST

A fungus that causes white pine blister rust, *Cronartium ribicola*, has a complex life cycle requiring two alternating hosts. It was first reported in Europe in 1854. Early workers found different life stages of the fungus growing on eastern white pine, which had been imported from the United States, and on currant and gooseberry plants (*Ribes* spp.). The fungus is believed to have been imported from eastern Asia, and to have spread by the movement of nursery stock throughout Europe. In 1910, an estimated 1,000 eastern white pine seedlings (some of which were diseased) were shipped from France to Vancouver, British Columbia. The fungus spread on *Ribes* and western white pine from British Columbia to New Mexico. Nine species of pines in the United States are at risk from this disease, which also infects many *Ribes* species.

Efforts to control white pine blister rust began in earnest with the passage of the Plant Quarantine Act of 1912. This law regulated the importation of plants or plant products capable of carrying pests or diseases. The first quarantine under the Act prohibited the importation of susceptible (5-needle) pines. Quarantine Number 26 authorized the destruction of wild and cultivated *Ribes* species. Thousands of young men were hired to uproot and eradicate *Ribes*. Some states offered landowners $.50 per plant to compensate for their losses, but most offered none. Emergency relief measures, the *Ribes* eradication program, and a controversial antibiotic treatment program over the next 50 years cost an estimated $50 million (16). Breeding programs have improved the resistance of western white pine seedlings, but losses of limber and whitebark pine still threaten fragile high-elevation ecosystems. Ancient bristlecone pines are also at risk.

PITCH CANKER

Many of the examples in this book describe destructive diseases brought into the United States, but the pathway is well-trodden in both directions. The pitch canker fungus, for example, is native to the Southeastern U.S., and perhaps Mexico. It infects pine trees through wounds and causes an unsightly, extensive pitch flow, but little mortality under normal conditions on its native hosts. In 1974, the same fungus caused a shoot dieback epidemic in Florida on planted slash pine (Figs. 2A & 2B). Even some of the normally less

Fig. 2. Shoot dieback (A) and copious resin flow (B) of pine canker. Photo courtesy of L. David Dwinell.

susceptible loblolly pines in seed orchards were affected. This unexplained, widespread outbreak has not recurred in the South. But the fungus began cropping up elsewhere.

In the 1980s, pitch canker was discovered in Japan, where at least three species of pine are known to be susceptible to the fungus. Four more species of Mexican pines were also added to the list of susceptible hosts. The disease has either become more widespread or better recognized in Mexico, where eight states have recently documented pitch canker occurrence.

In 1986, the fungus appeared suddenly in three separate locations along the coast of California, causing extensive branch dieback of Monterey pine. Natural resource professionals are concerned that we may entirely lose this central genetic source for Monterey pines, which is the most widely planted pine species in the world.

In 1990, a forest nursery in South Africa reported heavy losses of Mexican yellow pine (*P. patula*), which is widely planted there. Scientists found the fungus in South Africa was also attacking

41

Monterey pine in nurseries, but no outbreak occurred in that even more economically-important species in South Africa (10).

Seedlings of Monterey pine were found dying of pitch canker in the Basque Country in Northern Spain in 1997. And in the southeastern United States, the fungus has begun to cause 'damping-off' in longleaf and shortleaf pine seedlings. Many pine seeds have been found infected or coated with the fungus before planting. Scientists are now experimenting with modifying seed treatment practices to reduce seed contamination and the spread of pitch canker disease.

INK DISEASE

Symptoms of ink disease were observed on American chestnut between 1850 and 1875. There is evidence that the pathogen, *Phytophthora cinnamomi,* caused a devastating epidemic that was largely overshadowed by the later epidemic of chestnut blight. *Phytophthora cinnamomi* was first described in 1922 as the causal agent of stripe canker of cinnamon in Burma (12), and has since developed the reputation of being the single-most-cited plant pathogen in the world. At least 100 years previously, this microbe was transported around much of the world and encountered susceptible hosts wherever it was taken. It is still being introduced into new areas and finding new susceptible hosts. Well over 1,000 plant species (12), including several important forest tree species, are known hosts. It is a major contributing factor to littleleaf disease of shortleaf and loblolly pines in the southern United States (5), and is suspected to be contributing to oak decline in the southern Appalachian Mountains (13) and in Florida and Texas (23). On oaks and the very susceptible American chestnut, *P. cinnamomi* produces a distinctive basal trunk canker, which is surrounded by a thin, black zone. Hence, the name ink disease.

Other tree and understory plant species in many ecosystems are susceptible to infection by *P. cinnamomi*, but for unknown reasons, they are attacked and killed only under certain conditions. Plants that are stressed by excessively moist or dry soil conditions are sometimes more prone to infection (12). The pathogenic potential of *P. cinnamomi* is poorly understood because root pathogens are difficult to study and *P. cinnamomi* is difficult to isolate from infested soil and plant tissues. As this pathogen is transported into new areas, new potential hosts will be exposed to it and additional losses should be expected.

A little over a decade ago, *P. cinnamomi* was accidentally introduced into a dry native oak forest in southern Mexico. Since then, it has killed most of the oaks in a spreading infection center now over 300 hectares in size (22). It has been extremely difficult to assign blame for the decline because the affected trees have been attacked by a variety of root, stem, and bole fungi, as well as insects. This example of *P. cinnamomi* is used to illustrate the damage that one extremely devastating exotic pathogen can cause in a forested ecosystem. It might have been kept out of the country if we had known of its potential for destruction.

The involvement of *Phytophthora cinnamomi* in jarrah decline in western Australia is perhaps the best example of its destructive potential (11,19). Jarrah decline was first discovered in western Australia in 1921. Since then, this disease attacked more than 100 species of native flora, including the important timber species, jarrah (*Eucalyptus marginata*). It now threatens the ancient, species-rich heath communities of Australia (24) (Fig. 3). Most introduced plant pathogens in North America have affected only one or two host

Fig. 3. *Phytophthora cinnamomi* leaves few species surviving in this *Banksia* woodland in southwestern Australia.

genera, or species, but the experience to date in western Australia is a vivid example of the destructive nature of some introduced pathogens, especially if they can kill many host species.

Since 1995, a new species of *Phytophthora* has been causing a dramatic dieback of several coastal oak species in California. Unlike *P. cinnamomi*, this species attacks leaves, stems and branches rather than root systems. The pathogen, *P. ramorum*, causes minor leafspots on rhododendron and many other understory plants, but these appear to supply the inoculum by which the fungus spreads. These inconspicuous leafspots are causing major difficulties as regulators attempt to prevent the pathogen from being transported on nursery stock.

Yet another well-known example of a presumably exotic disease is dogwood anthracnose. Because this disease appeared suddenly, in epidemic proportions near Portland and New York City, decimating both eastern flowering dogwoods and Pacific dogwoods, while causing very insignificant leafspots on Chinese dogwood, it is thought that it may have reached our shores on nursery stock soon after trade reopened with China. But the origin of this fungus remains another unsolved mystery.

THE B STRAIN OF THE SWEET POTATO WHITEFLY: VECTOR FOR PLANT VIRUSES

Whiteflies are a group of insects that feed by penetrating the bark of tropical and subtropical plants with a structure called a stylet. In fact, they are not flies and are more closely related to mealybugs and scale insects (Fig. 4). Until 1957, 11 separate species of whiteflies were recognized based on descriptions in various parts of the world on various crop plants. At that time, all 11 were consolidated into a single species, *Bemisia tabaci*, because none of them could be visually distinguished from the original sweet potato whitefly (20).

In recent years, we have learned that even though whiteflies may look very much alike, their effects are very different. One in particular, the B strain, is causing major damage to irrigated crops throughout the dry tropics, where food supplies are often dangerously deficient.

Crop damage is caused by viruses that the whitefly carries from plant to plant. In Florida, it is blamed for a virus epidemic on tomato plants and a silvering in squash plants accompanied by a discoloration of the squash fruit. Moreover, the whitefly vector exhibited considerable resistance to widely used insecticides and demonstrated a remarkable ability for increasing its numbers rapidly.

Around the world, whiteflies were associated with damage to a wide variety of crops, including cassava in the Ivory Coast, beans in Puerto Rico, cucurbits (squash and melons), and cotton in various places. In many of the places where damage was observed, it became clear that the whiteflies differed in the hosts they attacked, even though they looked very much the same. Based on these observations, A and B types or strains of *B. tabaci* were proposed, as were a 'Sida' and a 'Jatropha' race.

Meanwhile, plant inspectors at ports of entry were unaware of these fine distinctions, and would not have been able to recognize them if they had known. Inspectors did know that whiteflies were already present in their countries, so they saw no need to intercept the ones that were present on imported vegetation. Their best efforts were not good enough because they lacked important knowledge. They had no idea how complex the world of insect vectors was. Without DNA analyses, we might never have realized that an exotic race of whiteflies, with a broad host range of several hundred species, wider geographic adaptation capacity, higher reproductive capacity,

Fig. 4. The sweet potato whitefly. Actual size = 0.8-1.2 mm.
Photo courtesy of R.C. Rosell, St. Thomas University, Houston, TX, and J.K. Brown, The University of Arizona, Tucson, AZ.

45

and more efficient in vectoring viruses, had displaced our native species (1,4,9).

The case of the sweet potato whitefly illustrates an intriguing philosophical dilemma that may have broad biological impact. Recent advances in biotechnology have permitted much deeper inquiry into the genetic differences among closely related organisms. We can now examine the genetic basis of relationships and detect differences between individuals, strains, and species that were not detectable based on morphology alone. We now know that for the sweet potato whitefly, strain level differences confer significant biological abilities relevant to quarantine officials. But in many other cases, the biological significance of small genetic differences are unknown. At this point, ignorance still plays a large part in decisions about how finely we divide species at international borders.

DUTCH ELM DISEASE

Most Americans and Europeans have witnessed, or at least heard about, the epidemic that destroyed many stately elms. In fact, in the last 80 years in the United States and Europe, hundreds of millions of these valuable forest, street, and landscape trees were lost (3). What most people do not realize is that there were really two epidemics, caused by different fungi.

No one knows where either fungus came from, but because the first pathogen, called *Ophiostoma ulmi*, has an optimum growth temperature of 22C, it is thought to come from a temperate climate, possibly Asia. Dying elms were first noticed in northwest Europe around 1910, and the epidemic swept across Europe and into North America around 1927, and into central Asia in the late 1930s. Only 10-40% of the European elms were lost, possibly because fungal viruses spread throughout that population. In the United States, elms were not so lucky. The virus has been found here, but is not very widespread.

Then, in the early 1970s, Europeans noticed a resurgence of the disease, and discovered that a different species of *Ophiostoma*, which was aptly named *O. novo-ulmi*, was causing the new epidemic. Retrospective studies found that the second epidemic had actually started in the 1940s. Unfortunately, these important facts were not known at the time, and elm logs were not prevented from international shipment. The disease worsened everywhere the new fungus was found.

Where the two fungi meet, *O. ulmi* is rapidly replaced by the more aggressive *O. novo-ulmi*. While both coexist, some interspecific hybridization occurs, and although the hybrids survive poorly, they serve as a genetic "bridge", permitting flow of pathogenicity genes and vegetative compatibility grouping genes. Having a diversity of vegetative compatibility groups reduces asexual mergers, and so reduces the spread of deleterious viruses within the pathogen population.

In addition, molecular studies have shown there are two different strains or subspecies of *O. novo-ulmi*, one that seems to have originated in Eastern Europe, and another that originated in the southern Great Lakes region of the United States. These interbreed freely in the laboratory. The spread of these strains has been tracked, and the ranges now overlap in several parts of Europe. Where they do overlap, recombinant forms are developing, suggesting that the adaptation of these fungi to new environments will continue. Recently, scientists looking for the geographic origin of these fungi discovered an entirely different species of *Ophiostoma* affecting elms in the Himalayas.

Dutch elm disease is a clear illustration of the need to avoid introducing genetic diversity into existing populations. Molecular geneticists need to work more closely with traditional biologists to define the biological basis for speciation. This information is sorely needed to guide future regulators in an important decision: Where do you draw the line?

Literature Cited

1.   Bedford, I.D., Briddon, RW., Brown, J.K., Rosell, R.C., Markham, P.G. 1994. Geminivirus transmission and biological characterization of *Bemisia tabaci* (Gennadius) biotypes from different geographic regions. Annals of Applied Biology 125:311-325.

2.   Brasier, C.M. 1996. *Phytophthora cinnamomi* and oak decline in southern Europe. Environmental constraints including climate change. Ann. Sci. For. 53:347-358.

3.   Braisier, C.M. 2001. Rapid evolution of introduced plant pathogens via interspecific hybridization. BioScience 5:123-133.

4.   Brown, J.K., Frohlich, D.R., and Roswell, R.C. 1995a. The sweetpotato or silverleaf whiteflies: Biotypes of *Bemisia tabaci* or a species complex? Annual Review of Entomology 40:511-534.

5.   Campbell, W.A., and Copeland, O.L., Jr. 1954. Littleleaf Disease of Shortleaf and Loblolly Pines. USDA Circ. 940. 41 p.

6.   Carefoot, G.L., and Sprott, E.R. 1967. "Famine on the Wind" Rand McNally & Co., NY. 229 pp.

7.   Chen, B., Choi, G.H. and Nuss, D.L. 1994. Attenuation of fungal virulence by synthetic infectious hypovirus transcripts. Science 264:1762-1764.

8.   Choi, G.H., and Nuss, D.L. 1992. Hypovirulence of chestnut blight fungus conferred by an infectious viral cDNA. Science 257:800-803.

9.   Costa, H.S., Brown, J.K., Sivasupramanium, S., and Bird, J. 1993. Regional distribution, insecticide resistance and reciprocal crosses between the A and B biotypes of *Bemisia tabaci*. Insect Science and Its Application 14:255-266.

10.   Dwinell, L.D., Adams, D., Guerra-Santos, J.J., and Aquirre, J.R.M. 1998. Pitch canker disease of *Pinus radiata*. Paper 3.7.30 *in*: Offered Papers Abstracts – Volume 3, Proc. 7th International Congress of Plant Pathology, Edenburg, Scotland, Aug. 6-16, 1998.

11.   Erwin, D.C., Bartnicki-Garcia, S., and Tsao, P.H. 1983. *Phytophthora* – Its Biology, Taxonomy, Ecology, and Pathology. APS Press, St. Paul, MN.

12.   Erwin, D.C., and Ribeiro, O.K. 1991. Phytophthora Diseases Worldwide. APS Press, St. Paul, MN.

13.   Jordan, A.P., and Tainter, F.H. 1996. The susceptibility of southern Appalachian oaks to *Phytophthora cinnamomi*. Castanea 61:348-355.

14. Kuhlman, E.G. 1978. The devastation of American chestnut by blight. Pp. 1-3 *in*: Proc. Am. Chestnut Symp., MacDonald, W.L., Cech, F.C., Luchok, J., and Smith, H.C. (eds.). West Virginia Univ. Books, Morgantown. 122 p.

15. Liebhold, A.M., MacDonald, W.L., Bergdahl, D., and Mastro, V.C. 1995. Invasion by Exotic Forest Pests: A threat to forest ecosystems. Forest Science. Monograph 30. 49 pp.

16. Maloy, O.C. 1997. White Pine Blister Rust Control in North America: A case history. Annu. Rev. Phytopathol. 35:87-109.

17. Mircetich, S.M., Campbell, R.N., and Matheron, M.E. 1997. Phytophthora trunk canker of coast live oak and cork oak in California. Plant. Dis. 61:66-70.

18. Mittempergher, L. 1978. The present status of chestnut blight in Italy. Pp. 34-37 *in*: Proc. Am. Chestnut Symp., MacDonald, W.L., Check, F.C., Luchok, J., and Smith, H.C. (eds.). West Virginia Univ. Books, Morgantown. 122 p.

19. Podger, F.D. 1972. *Phytophthora cinnamomi*, a cause of lethal disease of indigenous plant communities in Western Australia. Phytopathology 62:972-981.

20. Russell, L.M. 1957. Synonyms of *Bemisia tabaci* (Gennadius) (Homoptera: Aleyrodidae). Bulletin of the Brooklyn Entomological Society 52:122-123.

21. Shea, S.R., Shearer, B.L., Tippett, J.T., and Deegan, P.M. 1983. Distribution, reproduction, and movement of *Phytophthora cinnamomi* on sites highly conducive to jarrah dieback in southwestern Australia. Plant Dis. 67:970-973.

22. Tainter, F.H., O'Brien, J.G., Hernandez, A., Oroozco, F., Relolledo, O., and Appel, D.N. 1998. Oak decline in the state of Colima, Mexico (Abs.). Phytopathology 88:S88.

23. Van Arsdel, E.P., and Tainter, F.H. 1998. The role of Pythiaceous fungi in tree yellowing and death in Texas

(Abs.). Proc. 7th International Plant Pathology Congress
3:7.7.

24.	Wills, R.T. 1993. The ecological impact of *Phytophthora
cinnamomi* in the Stirling Range National Park, Western
Australia. Aust. J. Ecol. 18:145-159.

Plant-Parasitic Nematodes Which Are Exotic Pests in Agriculture and Forestry

L. David Dwinell
USDA Forest Service
Southern Research Station
Athens, GA

Paul S. Lehman
Florida Department of Agriculture and Consumer Services
Division of Plant Industry, Nematology Section
Gainesville, FL

Introduction

Insects, fungal pathogens, and weeds are infamous exotic pests; however, little is known about the effects of exotic plant-parasitic nematodes (Fig. 1). Most available information focuses on nematodes as quarantine pests and the regulatory process rather than their impact on invaded ecosystems. Published records of the interceptions at the borders of some countries give an impressive picture of the diversity of origin and the great number of species which are dispersed.

Nematodes, like other exotic pests, are not confined within geopolitical boundaries. The importance of nematodes as exotic pests may not have been fully realized because they are difficult to detect, and in some cases, it may be 10-50 years before infestation levels reach the point of causing recognizable damage. When a nematode suspected to be an exotic pest is discovered, it may take years of detective work and research to confirm its origin.

QUARANTINE PESTS

A quarantine pest is a noxious organism of potential economic importance to the area endangered and not yet present, or if present, it is not widely distributed and is being officially controlled (30). There are a number of plant-parasitic nematodes of regulatory concern. Potato cyst nematodes (*Globodera rostochiensis* and *G. pallida*) are prohibited by plant health legislation in the United States, Canada, Australia, New Zealand, Mexico, Chile, and Japan (5,25).

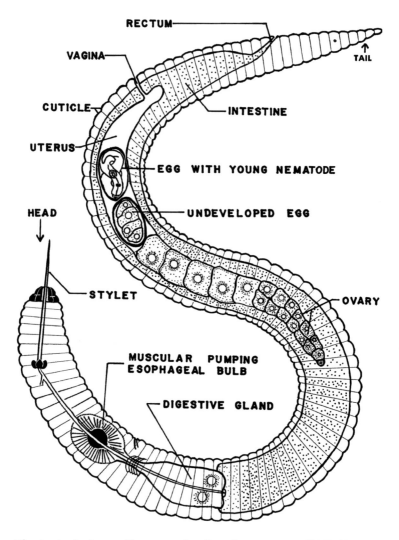

RECTUM

VAGINA

CUTICLE

UTERUS

HEAD

TAIL

INTESTINE

EGG WITH YOUNG NEMATODE

UNDEVELOPED EGG

STYLET

OVARY

MUSCULAR PUMPING
ESOPHAGEAL BULB

DIGESTIVE GLAND

Fig. 1. A plant-parasitic nematode. Drawing courtesy of R.P. Esser.

The European Union also regulates *Longidorus diatecturus* (needle nematode), *Nacobbus aberrans* (*sensu lato*), *Ditylenchus destructor* (stem nematode), *D. dipsaci, Aphelenchoides besseyi* (bud and leaf nematode), *Radopholus similis* (burrowing nematode), *R. citrophilus,* and *Bursaphelenchus xylophilus* (pinewood nematode) (5,38). The European and Mediterranean Plant Protection Organization's A1 list of quarantine nematodes includes those of the European Union as well as *Heterodera glycines* (soybean nematode) and *Xiphinema americanum sensu lato* (dagger nematode) (37). The principal nematodes of regulatory concern among North American Plant Protection Organization countries include *Globodera rostochiensis, G. pallida, Meloidogyne chitwoodii* (root-knot nematode), *Heterodera trifolii* (clover cyst nematode), *H. zeae* (corn cyst nematode), *H. goettingiana* Liebscher (pea cyst nematode), *H. glycines, Ditylenchus dipsaci* and *Radopholus similis* (25).

Quarantine pests are potential exotic pests. In order to exclude plant pests, countries have established regulatory agencies. In the United States, the agency responsible for the national effort to exclude plant pests is the Department of Agriculture's Animal and Plant Health Inspection Service (APHIS). Between 1985 and 1997, APHIS intercepted plant parasitic nematodes 200 times at U.S. ports (J. Cavey, personal communication). Agriculture and Agri-Food Canada reported 501 interceptions of nematodes for 1991-1994 (1).

Fortunately, very few exotic nematode pests become established. An exotic pest will establish only when a susceptible host is present, the parasite is virulent and capable of reproducing, and the environment is favorable (21,28). This interaction of host, pathogen, and environment is conceptualized in what plant pathologists call the Disease Triangle.

HOST-PATHOGEN SYSTEMS

Two basic host-pathogen models predict the consequences of introducing exotic pests into a new environment. The natural resources model is the most frequently cited (Table 1). The basis of this model is the principle that when the host-pathogen system has co-evolved, it has usually reached a state of equilibrium in which the parasites do not exterminate their hosts and the hosts have not developed sufficiently effective defenses to eliminate the pathogen (35). If, instead of co-evolving, a pest is introduced into a host population that lacks defenses against it, the damage may be disastrous. The pine wilt disease in Japan caused by a nematode introduced from North America, *Bursaphelenchus xylophilus,* is an example that fits this model (8,23,27).

Since nearly all of the agronomic and horticultural plants in North America are introduced, the agricultural model (Table 2) is applicable. The agricultural model, however, has been over-shadowed by the

Table 1. Natural resources model

Host	Pathogen	Disease
Native	Native	Low
Native	Introduced	High

Table 2. Agricultural model

Host	Pathogen	Disease
Introduced	Native	Low
Introduced	Introduced	Variable

natural resources model. As Yarwood (47) points out a major reason crops do better in areas remote from their place of origin is that when they are introduced, their pathogens, or some of them, are left behind. Many times, however, the pathogens enter with their hosts or are reunited with their long-departed hosts (34,35,47). The consequences of this reunion depend largely on how long the host has been under domestication, whether or not it has lost its resistance to the pathogen, and whether, during the interim, the pathogen's virulence has changed significantly. The reuniting of the potato cyst nematode and the soybean cyst nematode with their hosts which were also introduced into the United States, represents the agricultural model.

EXOTIC NEMATODE CASE STUDIES

Potato Cyst Nematodes

Currently two species of cyst-forming nematodes, *Globodera rostochiensis* and *G. pallida,* are major pests of potatoes (*Solanum* spp.). For several decades, the center of origin of the two species, where they had co-evolved with their preferred host, the potato, was thought to be the high Andes of South America. Most recently, however, ribosomal DNA sequence data has led to the thesis that they evolved in Mexico and may have migrated on wild *Solanum* spp. southward to the Andes (13). Potatoes were introduced to Spain during the last half of the 16th century (3,4). Potato cyst nematodes probably did not reach Europe until the mid 19th century in potato collections brought to breed for resistance to a fungal disease called late blight (3). The nematodes spread with seed potatoes (5). In North America, *Globodera rostochiensis,* the golden nematode, was reunited with potatoes in Canada, Mexico, and the United States. *Globodera pallida* is known to occur in Canada (25), but not yet in the United States. So, although host and pathogen were both introduced, after four centuries of domestication potatoes had essentially lost their resistance to cyst nematodes.

In the United States, the golden nematode was first detected in 1941 on Long Island in New York State. It was perhaps introduced on military equipment returning from Europe after World War II (2). The original infested field was part of an airfield of a temporary military camp. The nematode subsequently spread to 30 additional fields farmed by the operator of the original infested field (3).

Soon after detection, a New York regulatory agency took action to prevent the spread of this exotic pathogen. In 1944, the State implemented specific quarantine regulations. In 1948, the federal government enacted the Golden Nematode Act to protect the potato and tomato industries from the golden nematode. Seed certification programs were also established. Through highly cooperative efforts of state and federal research, regulatory, and extension personnel, as well as potato growers, the infestation has been confined to two counties on Long Island and four other counties in western New York.

Soybean Cyst Nematode

The soybean cyst nematode, *Heterodera glycines,* illustrates how difficult it can be to determine whether a nematode is an exotic pest. Soybean (*Glycines max*) originated in China, and was introduced in Japan and Korea between 200 B.C. and 300 A.D. Soybean seeds were imported into the United States from China in the late 1700s (31,33).

In 1954, the soybean cyst nematode was found in North Carolina. It is likely, however, that the pest was introduced at a number of U.S.

locations earlier. Noel (31) hypothesizes that it co-evolved with soybeans in Asia and was widely dispersed in the United States in the late 1890s. At that time, soil was imported from the Orient to inoculate soybeans with the nitrogen-fixing *Bradyrhizobium japonicum*. The drawbacks to such transfers of soil from one field to another were first noted by Temple in 1916 (31).

Soybeans were not a major crop in the United States until World War II. At that time, soybean production increased significantly to offset losses in vegetable oil. The nematode became a problem after the advent of modern agriculture and the abandonment of rotations in which soybeans were planted every 3 or 4 years. By 1960, established infestations were discovered in North Carolina, Tennessee, Missouri, Arkansas, Illinois, Mississippi, Kentucky, and Virginia. The soybean cyst nematode is now found in 25 states (31,37).

The soybean cyst nematode was found in Brazil in 1991. By 1995, losses in Brazil due to this exotic pest were estimated to be 216,000 metric tons valued at $32 million US dollars (15).

Pinewood Nematode

The pinewood nematode (*Bursaphelenchus xylophilus*) is native to North America where the damage it causes is negligible (8). In the early 1900s, when this nematode reached Japan, it caused rapid wilting of *Pinus thunbergii,* Japanese black pine, and *P. densiflora,* Japanese red pine (9,23,26,27). It was not until 1971 that the pinewood nematode was recognized as the causal agent (24), and North America was identified as the source (27). The pinewood nematode was probably spread from Japan to Taiwan, the People's Republic of China, and South Korea (27). In 1999, the pinewood nematode was reported in Portugal (29), the first report of its occurrence in Europe. It is considered to be an exotic pest in these countries (27,29), and its behavior is certainly following the natural resources model.

Most of the 49 described species of *Bursaphelenchus* are carried by insects, especially bark beetles and woodborers. The pinewood nematode and other closely related species of *Bursaphelenchus* (such as *B. mucronatus*) are vectored principally by cerambycid longhorn beetles (also called sawyers) in the genus *Monochamus.* The biologies of the Asian, North American, and Euro-Siberian *Monochamus* species are similar (9). Adult sawyers are attracted to recently dead or dying trees and freshly felled timber (including logs) for breeding. During insect oviposition, the nematode enters into the oviposition wounds, then invades the plant tissue and feeds largely on fungi in the wood (9,45). This is called secondary transmission because the nematode is not the primary cause of death. However, the nematode can move across international borders in logs infested in this manner (8,9,45).

After emerging from the log, the adult beetles feed on the bark of young branches, transmitting the nematode through wounds to live pines. The nematode reproduces rapidly and causes the infected tree to show symptoms of decline and wilt. In such events, the pinewood nematode is functioning as a primary pathogen and the resulting disease is pine wilt (8,23,27,45).

It is highly likely that the pinewood nematode entered Japan in pine logs later colonized by the native sawyer (*Monochamus alternatus*) and that *B. xylophilus* became adapted to the Japanese pine sawyer and displaced the native nematode *B. mucronatus,* which did no damage to Japanese pine species. Essentially, in countries where pine wilt disease is epidemic, *B. xylophilus* is an exotic pest, being vectored by native insects, *Monochamus* spp., which are components of the pine ecosystem. This combination of factors would make the pine wilt disease almost impossible to control.

Burrowing Nematode

Some races of nematodes that cause significant economic losses are considered exotic and excluded by many countries, even though other races of the same species may already be present. Because races cannot be distinguished morphologically, regulatory agencies generally exclude them all.

Burrowing nematode, *Radopholus similis sensu lato,* is a banana pest which has been widely disseminated in tropical and subtropical regions of the world via infected banana corms and other vegetatively propagated crops (7). In the early 1950s, the citrus industry in the United States was threatened by a devastating disease, known as spreading decline, that was discovered to be caused by burrowing nematode. The first evidence of physiological races of *Radopholus similis* came from field observations of banana growing adjacent to citrus in Florida (7). Greenhouse studies showed that some burrowing nematodes can parasitize bananas, but not citrus. These were designated as the banana race. Others, which infect both citrus and banana, were designated as the citrus race. It soon became evident that the burrowing nematode causing spreading decline of citrus in Florida did not occur in other major citrus-growing areas of the world.

Other U.S. states and other countries where citrus is a major crop developed regulations to prevent the introduction of the citrus race of the burrowing nematode which causes serious losses to citrus in Florida (25).

Citrus Nematode

Citrus nematode, *Tylenchulus semipenetrans,* occurs in all major citrus-producing countries of the world, and by the strictest definition, would not be classified as a quarantine pest in any of these countries. This nematode pest, however, illustrates the need to rethink pest exclusion policies that are based primarily on organism distribution in relation to geopolitical boundaries. Even if this nematode already occurs within a country, citrus nematode biology suggests that it should be considered regionally exotic and excluded from any area where sampling indicates the absence of citrus nematode.

The citrus nematode probably co-evolved with citrus or citrus relatives in southeast Asia. Most likely, it spread around the world on citrus rootstocks as they were introduced into new regions. This nematode is a very specialized parasite, in that it develops high densities on its obligate host with relatively low damage, and it has a narrow host range. It may be disseminated on other crop plants such as olives, persimmons, and grapes – but it has few other reservoir hosts. In Florida, it was first thought to occur on weeds and native plants, but it is now known that these nematodes are other species of *Tylenchulus,* which do not parasitize citrus (19). Even within infested groves or citrus orchards, citrus-nematode distribution and density are generally not very uniform, reflecting that this nematode has evolved a very specialized form of parasitism that has minimized the selection pressure for specialized survival and dispersal modes outside its host. Therefore, the citrus nematode can be excluded from very limited areas, if sanitation practices are established for groves and nurseries.

REGULATORY PROCESS AND TRADE

Since quarantine procedures are exclusionary, they frequently disrupt international trade. Pest risk assessments help define the risks/benefits of a pest by determining whether a pest is a quarantine pest and by evaluating its introduction potential (30). If the biological characteristics of an exotic pest are not fully known, the concept that is frequently followed is "when in doubt, keep it out" (21).

Unfortunately, there is a paucity of data on the impact of quarantine procedures on trade. Regulatory issues and trade issues are often interlinked. Nematodes have played an important role in the regulatory process in North America, Europe, and elsewhere. Some examples follow.

Japanese Flowering Cherry

In 1908, Mrs. William Howard Taft wanted to beautify the corridor that stretches from the presidential memorials to the capital. It was David Fairchild, who had established the Office of Foreign Seed and Plant Introduction, who convinced the President's wife that Japanese flowering cherry trees should grace the area. Before long, the news arrived through diplomatic channels that the mayor of Tokyo would donate 2,000 trees to the project. The trees arrived from Japan in Seattle on December 10, 1909, and were sent by train to Washington. When the trees arrived in early January, the Department of Agriculture's inspection team, which included the eminent nematologist N.A. Cobb, examined them and discovered several types of scales, wood-boring larvae, and galls caused by root-knot nematodes. In their report to the Secretary of Agriculture, the head of the inspection team stated that the shipment had the "worst infestation of insects and root galls he had ever encountered" and courageously recommended that the trees be "destroyed by burning". In the final report, Cobb stated "I have no hesitation in saying that in a country where a proper inspection of diseased material was legally in force with the objective of protecting agriculture, the importation of these trees would not be permitted." The following year, the Japanese government sent a new pest-free gift of 6,000 trees that had been fumigated with hydrocyanic acid gas to assure that the embarrassing incident would not be repeated. (20). These events along with introductions of the boll weevil, white pine blister rust, and chestnut blight, led to the passage of the Federal Quarantine Act of 1912 (20,47).

Siberian Larch Logs

In 1990, there was interest in importing logs from Siberia to offset timber supply problems in California caused by the endangered northern spotted owl. However, a test shipment of pine logs (*Pinus sylvestris*) was found to be infested by *Bursaphelenchus mucronatus* and other pests (39). The headline of an article in The Journal of

Commerce read "A Soviet Pine Penned – Tiny Worm Bugs Up Test Deal" (July 5, 1990). This nematode, which is common in Asia and Europe, was not known in the United States so the concept of "when in doubt, keep it out" came in-to play. However, of the 49 described species of *Bursaphelenchus,* only two species – *B. xylophilus* and *B. cocophilus* (the red ring nematode of palms) are generally considered to be pathogens (8,9,23). As a result of the fallout from this interception, APHIS requested that the USDA Forest Service provide a pest risk assessment on Siberian timber (43). The pest risk assessment concluded that "The pine wood nematode would affect ponderosa pine stands in California and the Southwest, converting them to scrub or nonforest. At least one animal, the tassel-eared squirrel, would be endangered, and the regional hydrology would be significantly altered." Although *B. mucronatus* has been recovered from dead or dying exotic or native pines where the nematode is known (23), there is little evidence that the nematode killed the trees. More often than not, the nematode is a secondary associate, and not the cause of mortality (8,9). These events precipitated the USDA/APHIS regulations governing the importation of logs, lumber, and other manufactured articles. The Final Rule was published in 1995 (41). In June 1997, a federal judge, in response to a lawsuit brought by a coalition of environmental groups (Oregon Natural Resources Council vs. Animal and Plant Health Inspection Service), issued a preliminary injunction limiting imports of raw logs and other wood products into the United States because of concerns that they may carry destructive pests (42). In January 1999, the U.S. District Court for the Northern District of California removed the court-imposed injunction on the issuance of new wood import permits. The forest industries in the Pacific Northwest, however, have not been able to turn to the forests of Asiatic Russia for unprocessed coniferous wood. Now, however, raw radiata pine (*Pinus radiata*) lumber from Chile and New Zealand and raw Douglas-fir (*Pseudotsuga menziesii*) lumber from New Zealand may enter the United States under written permit.

Softwood Exports to Europe

The American pinewood nematode, *B. xylophilus,* has also influenced the regulatory process in Europe. When Finnish inspectors intercepted the pinewood nematode in imported pine chips from North America in 1984, a ban was placed on the import of conifer chips and timber cut from softwood trees grown in North

America, China, Japan, Taiwan, and South Korea (32). Sweden and Norway imposed similar import restrictions. In July 1985, the European and Mediterranean Plant Protection Organization placed the pinewood nematode on its A1 list of quarantine pests for Europe. It recommended that its member countries ban soft-wood products (except kiln-dried lumber) from countries known to have *B. xylophilus* (36). By 1989, the European Union had emerged as the most visible regulatory entity in Europe (8). The issue escalated in the early 1990s when Finnish inspectors intercepted the pinewood nematode in unseasoned lumber (40) and packing-case wood (39) imported from Canada. These concerns lead to trilateral research involving scientists in Canada, the United States, and the European Union on the efficacy of heat-treating unseasoned coniferous lumber to eradicate the pinewood nematode and its vectors (10). The European Union now requires pasteurization or kiln-drying of coniferous wood before importation (8).

Because of the ban, no pine chips have been exported from North America to Europe since 1986. Forest industry in the southern United States estimates a $20 million loss per year in potential sales of pine chips to European markets as a direct result of the embargoes. Annual losses in green lumber exports are estimated to be more than $100 million annually in the United States (8). The value of Canada's losses is considerably higher (8).

Wooden Packing Material

One entry pathway for exotic pests is in wooden packing material (i.e., pallets, crating, and cable spools) or dunnage used to ship cargo. This issue came to the forefront with the discovery of established populations of the Asian longhorned beetle (*Anoplophora glabripennis*) in the United States in 1996 (17). In 1991, Tomminen (39) had reported recovering the pinewood nematode from packing-case wood. In China, the pinewood nematode was intercepted in 1992 in wood packing and supporting material of a sightseeing wheel imported from Japan (46). The European Union and the People's Republic of China now regulate coniferous wood packing material to minimize the risk of additional introductions of the pinewood nematode and its *Monochamus* vectors. Concerns about pest movement in wooden packing material were the subject of a 2001 international on-line workshop (www.exoticpests.apsnet.org). A CD-ROM of the workshop is also available from the USDA Forest Service, Southern Research Station. In 2000, the United States

Department of Agriculture published a draft pest risk assessment for importation of solid wood packing material for public comment, and in 2003 is moving to harmonize mitigation requirements for wooden packing material with standards proposed by the International Plant Protection Convention (16).

Citrus Burrowing Nematode and Ornamentals

Regulatory action to exclude citrus burrowing nematode affects not only the movement of citrus hosts in international trade, but also impacts the ornamental plant industry in Florida, Hawaii, and other tropical countries. Host studies on the citrus race of the burrowing nematode indicated that it reproduces on ornamental plants in diverse plant families. Many of these ornamental plants are hosts of the banana race of the burrowing nematode. Citrus-producing states and countries also have imposed regulatory restrictions on any burrowing nematode associated with ornamental plants exported from Florida, because they cannot distinguish the banana race from the exotic citrus race using routine morphometric identification techniques.

In 1984, the citrus race of the burrowing nematode was described as a new species, *Radopholus citrophilus* (18). Since the newly described species did not differ morphologically from *R. similis* the two were separated as sibling species. Although this taxonomic change was intended to assist regulatory agencies, it has resulted in some confusion. Some countries, such as Japan and the European Union, have modified their regulations to prohibit both sibling species, whereas others continue to prohibit only *R. similis sensu lato,* but include sibling species under one species, because morphologically they are indistinguishable. From a practical standpoint, however, regulatory agencies continue to exclude burrowing nematodes that parasitize citrus, since they either prohibit both races or both sibling species. In 1998, Valette et al. (44) proposed, based on scanning electron microscopic studies comparing *R. similis* populations from Africa, that the two species should be synonomized, with *R. citrophilus* as a junior synonym of *R. similis.* This taxonomic revision is supported by independent molecular and biochemical studies (22) and morphological studies of eight *R. similis* populations from three continents (11). These changes in taxonomic nomenclature illustrate that it is important for regulatory agencies to understand the lineage of pest names and revise their regulations as taxonomic names are changed in order to prevent confusion as to which pests are regulated in international trade.

COST AND BENEFITS OF EXCLUDING NEMATODES

Regulatory programs in the USA have reduced the spread of economically important nematodes such as potato cyst nematodes, the burrowing nematode, the citrus nematode, and the reniform nematode, *Rotylenchulus reniformis.*

Since the golden nematode was detected about 50 years ago, federal regulatory programs have helped to prevent its spread. In 1995, the value of the U.S. potato crop was $3 billion. In many European countries where this nematode is widespread, losses are 8 to 10%. The benefit of protecting the U.S. potato crop from a loss of 9% would be $270 million, far exceeding the federal regulatory budget for potato cyst nematode exclusion programs, which in 1996 was $445,000 (25).

For the past 40 years, regulatory programs in Florida have effectively limited the spread of the burrowing nematode in citrus. Considering the losses caused by *R. similis* and on-tree values of citrus during the past 35 years converted to 1995 dollar values, the $1.4 billion cumulative benefits to Florida citrus growers far outweighed the $100 million spent excluding burrowing nematodes from 18,000 hectares.

Nursery certification programs for burrowing nematode also have excluded citrus nematode from citrus seedlings and from new planting areas at a $32.5 million benefit to Florida growers. Other state regulatory programs exclude economically important nematodes found in Florida and other states. An example is reniform nematode, which is not found on cotton in the states of California and Arizona. Cooperative interstate ornamental nursery sanitation and regulatory certification programs were implemented in Florida, Hawaii, Louisiana, Texas, and Puerto Rico to reduce the risk of introducing this nematode. The benefits of excluding reniform nematodes from cotton in California in 1995 was estimated to be $8.5 million. Regulatory programs that exclude nematodes have been extremely cost-effective in the U.S.

FUTURE NEEDS

This past decade has been marked with numerous free trade agreements that have resulted in the emergence of multinational trading blocks. As regional trading blocks merge and expand in the future, we will need to shift both our way of thinking and our regulatory policies to meet the challenges of exotic pests.

Regulatory agencies must move from pest exclusion policies that have been developed along geopolitical lines to policies that are multinational and based on pest biology and ecology. This need was recognized when the International Plant Protection Convention (IPPC) was established in 1951 as a subsidiary of the United Nations Food and Agricultural Organization (FAO). There are more than 100 signatory nations to the IPPC agreement, which provides a framework for regional and global efforts in plant protection.

Efforts to exclude exotic pests are as interdependent as links in a chain. The effectiveness of efforts to prevent nematode introduction is limited by the weakest exclusionary efforts. In general, failure to exclude exotic nematodes can be traced either to policies that are based on incomplete information about pest biology or to inconsistent application of biological knowledge because of political or economic considerations. Attempts to limit the spread of soybean cyst nematode in the United States illustrate how incomplete biological knowledge can result in ineffective regulatory efforts. In this case, ground transportation at inspection stations was shown to have limited value since birds drop viable soybean cyst nematodes as they fly overhead. Many exclusionary procedures for exotic nematodes need to be strengthened based on current biological knowledge.

Quarantine lists often are not well balanced and appropriately updated. In Europe, for example, quarantine lists tend to emphasize exotic nematodes that are known to damage crops in temperate climates. EPPO lists of exotic pests that warrant exclusionary efforts may need to include nematodes that likely have potential to become established and damage subtropical crops in southern Europe. Examples of such nematodes are *Rotylenchulus reniformis* (burrowing nematode), *Pratylenchus coffeae* (lesion nematode), *Belonolaimus* (sting nematode), and *Hoplolaimus* (lance nematode) species.

Another weak link is the rapid change in patterns of commerce and trade that has influenced the ornamental plant industry worldwide during the past 15 years. Many countries, including the United States, for many years have prohibited the importation of soil or plants in soil, but allowed the importation of plant cuttings and bare-root plants. This policy was established when few bare-rooted plants were imported and when it was thought that most root pests could be detected by visual inspection. To reduce labor cost, more and more nurseries have multi-national operations and air-freight planting material from other countries. For example, ornamental nurseries in Florida import about 500 million cuttings and bare-rooted plants annually from more than 40 different countries. However, scientists know that plant-parasitic nematodes can migrate into stem tissue or remain attached to minute roots, and that ectoparasitic nematodes are introduced with soil adhering to bare-rooted plants. If regulatory policy does not keep pace with current market practices and new understanding of nematode biology, ornamental plants could serve as a "Trojan horse" for introducing exotic nematodes to food crops notwithstanding stronger sanitation requirements that prohibit importation of plants in soil. Federal inspection standards at the ports of entry or sanitation standards at the site where bare-rooted plants and cuttings originated need to be modified to strengthen this weak link to exclude exotic nematodes.

Conclusion

In coming decades, the expanding world population will increase pressures for food and fiber, and the desire to share in the benefits of technology will continue to spread globally. As these trends conflict with Earth's fragile ecosystems and global environmental concerns increase, there will be a growing need for cooperative regional biologically-based regulatory policy. The benefits from excluding exotic nematode pests will become even greater in the coming decades.

Literature Cited

1. Agriculture and Agri-Food Canada. 1994. Intercepted plant pests. Center of Expertise for Quarantine Pests, Animal and Plant Health Directorate, Food Production and Inspection Branch, Agriculture and Agri-food Canada, Nepean, Ontario, Canada. 119 pp.

2. Brodie, B.B. 1984. Nematode parasites of potato. Pages 167-212 *in*: Nickle, W.R., ed. Plant and insect nematodes. Marcel Decker, New York. 925 pp.

3. Brodie, B.B., and Mai, W.F. 1989. Control of the golden nematode in the United States. Annu. Rev. Phytopathol. 27:443-461.

4. Brodie, B.B., Evans, K., and Franco, J. 1993. Nematode parasites of potatoes. Pages 87-132 *in*: Evans, K., Trudgill, D.L., and Webster, J.M., eds. Plant parasitic nematodes in

temperate agriculture. CAB International, Wallingford, UK. 648 pp.

5.   Cotten, J., and Van Riel, H. 1993. Quarantine: problems and solutions. Pages 593-607 *in*: Evans, K., Trudgill, D.L., and Webster, J.M., eds. Plant parasitic nematodes in temperate agriculture. CAB International, Wallingford, UK. 648 pp.

6.   Crawford, R.P. 1922. World crops of America. Sci. Amer. 126:226-227.

7.   DuCharme, E.P., and Birchfield, W. 1956. Physiological races of the burrowing nematode. Phytopathology 46:615-616.

8.   Dwinell, L.D. 1997. The pinewood nematode: regulation and mitigation. Ann. Rev. Phytopathol. 35:153-166.

9.   Dwinell, L.D., and Nickle, W.R. 1989. An overview of the pinewood nematode ban in North America. Gen. Tech. Rep. SE-55, USDA For. Serv., Southeast. For. Exp. Sta., Asheville, NC. 13 pp.

10. EOLAS, Irish Science and Technology Agency. 1991. The development of heat treatment schedules to ensure eradication in lumber of the pinewood nematode (*Bursaphelenchus xylophilus*) and its insect vectors. Final report., EOLAS, Glasnevin, Dublin.

11. Elbadari, G.A., Geraert, E., and Moens. 1999. Morphological differences among *Radopholus similes* (Cobb, 1893) Thorne, 1949 populations. Russian J. Nematol. 7:139-153.

12. Fairchild, D. 1906. Our plant immigrants. Natl. Geogr. Mag. 17:179-291.

13. Ferris, V.R., Miller, L.L., Faghihi, J., and Ferris, J.M. 1995. Ribosomal DNA comparisons of Globodera from two continents. J. Nematol. 27:273-283.

14. Foster, J.A. 1991. Exclusion of plant pests by inspections, certification and quarantines. Pages 311-338 *in*: Hanson, A.A., and Pimental, D., eds. CRC Handbook of Pest Management in Agriculture, 2nd edition, Vol. 1. CRC Press, Boca Raton, FL.

15. Goellner, G.F. 1995. Aspectos economicos do nematoide do cisto na sojacultura Brasileira. Pages 102-106 *in*: Program and Proceedings of Congresso International de Nematogia Tropical, June 4-9, 1995, Rio Quente, Brasil.

16. Griffin, R.L. 2000. Forest biosecurity and the international harmonization of phytosanitary measures for wood products. Proc. XIV-SILVTECNA Quarantine Pests, Risks for the Forestry Sector and Their Effects on Foreign Trade, 27-28 June, 2000, Concepcion, Chile. 19 pp.

17. Haack, R.A., Law, K.R., Mastro, V.C., Ossenbruggen, H.S., and Raimo, B.J. 1997. New York's battle with the Asian long-horned beetle. J. Forestry 95:11-15.

18. Huettel, R.N., Dickson, D.W., and Kaplan, D.T. 1984. *Radopholus citrophilus* sp. n. (Nematoda), a sibling of species of *Radopholus similis*. Proceedings Helminthological Society of Washington 51:32-35.

19. Inserra, R.N., Vovlas, N., O'Bannon, J.H., and Esser, R.P. 1988. *Tylenchulus graminis* n. sp. and *T. palustris* n. sp. (Tylenchulidae), from native flora of Florida, with notes on *T. semipenetrans* and *T. furcus*. J. Nematol. 20:266-287.

20. Jefferson, R.M., and Fusonie, A.E. 1977. The Japanese cherry trees in Washington, D.C. National Arboretum Contribution No. 4. 66 pp.

21. Kahn, R.P. 1989. Quarantine significance: biological considerations. Pages 201-216 *in*: Plant Protection and Quarantine, Vol. 1, Biological Concepts. CRC Press, Boca Raton, FL. 226 pp.

22. Kaplan, D.T., Vanderpool, M.C., and Opperman, C.H. 1997. Sequence tag site and host range assays demonstrate

that *Radopholus similis* and *R. citrophilus* are not reproductively isolated. J. Nematol. 29:421-429.

23.   Kishi, Y. 1995. The pine wood nematode and the Japanese pine sawyer. Tokyo, Japan: Thomas. 301 pp.

24.   Kiyohara, T., and Tokushige, Y. 1971. Inoculation experiments of a nematode, *Bursaphelenchus* sp. onto pine trees. J. Jap. For. Soc. 53:210-218.

25.   Lehman, P.S. 1995. Role of plant protection organizations in nematode management. Pages 137-148 *in*: Proc. Congresso International de Nematologia Tropical, June 4-9, 1995, Rio Quente, Brasil.

26.   Mamiya, Y. 1983. The pathology of the pine wilt disease caused by *Bursaphelenchus xylophilus*. Annu. Rev. Phytopathol. 21:201-220.

27.   Mamiya, Y. 1987. Origin of the pine wood nematode and its distribution outside the United States. Pages 59-65 *in*: Wingfield, M.J., ed. Pathogenicity of the pine wood nematode. APS Press, St. Paul, MN. 122 pp.

28.   Mass, P.W. 1987. Physical methods and quarantine. Pages 265-291 *in*: Principles and practices of nematode control in crops. Academic Press, Australia.

29.   Mota, M.M., Braasch, H., Bravo, M.A., Penas, A.C., Burgermeister, W., Metge, K. and Sousa, S. 1999. First report of *Bursaphelenchus xylophilus* in Portugal and in Europe. Nematology 1:727-734.

30.   NAPPO. 1995. NAPPO Compendium of Phytosanitary Terms. Secretariat, Nepean, Ontario, Canada. 21 pp.

31.   Noel, G.R. 1992. History, distribution, and economics. Pages 1-13 *in*: Riggs, R.D. and Wather, J.A., eds. Biology and management of the soybean cyst nematode. APS Press, St. Paul, MN.

32.   Rautapaa, J. 1986. Experiences with *Bursaphelenchus* in Finland. Conf. Pest and Disease Problems in European Forests. EPPO Bull. 16:453-456.

33.   Riggs, R.D., and Niblack, T.L. 1993. Nematode pests of oilseed crops and grain legumes. Pages 209-258 *in*: Evans, K I., Trudgill, D.L., and Webster, J.M., eds. Plant parasitic nematodes in temperate agriculture. CAB International, Wallingford, UK. 648 pp

34.   Schoulties, C.L., Seymour, C.P., and Miller, J.W. 1983. Where are the exotic disease threats? Pages 139-181 *in*: Wilson, C.L., and Graham, C.L., eds. Exotic plant pests and North American Agriculture. Academic Press, New York. 522 pp.

35.   Simons, M.D., and Browning, J.A. 1983. Buying insurance against exotic plant pathogens. Pages 449-478 *in*: Wilson, C.L. and Graham, C.L., eds. Exotic plant pests and North American Agriculture. Academic Press, New York. 522 pp.

36.   Smith, I.M. 1985. Pests and disease problems in European forests. FAO Plant Prot. Bull. 33:159-64.

37.   Smith, I.M., McNamara, D.G., Scott, P.R., and Harris, K.M., eds. 1992. Quarantine pests for Europe. CAB International, Wallingford, UK. 1032 pp.

38.   Tacconi, R., and Ambrogioni, L. 1995. Nematodi da Quarantena. Lo Scarabeo, Bologna. 191 pp.

39.   Tomminen, J. 1991. Pinewood nematode, *Bursaphelenchus xylophilus,* found in packing case wood. Silva Fenn. 25:109-111.

40.   Tomminen, J., and Nuorteva, M. 1992. Pinewood nematode, *Bursaphelenchus xylophilus,* in commercial sawn wood and its control by kiln-heating. Scand. J. For. Res. 7:113-120.

41.   United States Department of Agriculture, Animal and Plant Health Serv. 1995. Importation of logs, lumber, and other unmanufactured articles. Final Rule. Federal Register 60:27665-82.

42.   United States Department of Agriculture, Animal and Plant Health Serv. 1997. Supplemental environmental impact statement for the importation of logs, lumber, and other unmanufactured wood articles. Federal Register 62:45217.

43.   United States Department of Agriculture, For. Serv. 1991. Pest risk assessment of the importation of larch from Siberia and the Soviet Far East. Pub. No. 1495.

44.   Valette, C., Mounport, D., Nicole, M., Sarah, J., and Baujard, P. 1998. Scanning electron microscope study of two African populations of *Radopholus similis* (Nematoda: Pratylenchidae) and proposal of *R. citrophilus* as a junior synonym of *R. similis.* Fundam. Appl. Nematol. 21:139-146.

45.   Wingfield, M.J., Blanchette, R.A., and Nicholls, T.H. 1984. Is the pinewood nematode an important pathogen in the United States? J. For. 82:232-235.

46.   Xu, P., Ming, T., Peiyin, S., and Guoyao. 1995. The interception of pinewood nematode on the wooden packing and supporting material of sightseeing-wheel imported from Japan. International Symposium of Pine Wilt Disease, 31 October–5 November 1995, Beijing, China.

47.   Yarwood, C.E. 1983. History of plant pathogen introductions. Pages 39-63 *in*: Wilson, C.L., and Graham, C.L. eds. Exotic plant pests and North America Agriculture. Academic Press, New York. 522 pp.

Meeting the Threat:
Risk Assessment and Quarantine

Matthew H. Royer and Ed Podleckis
U.S. Department of Agriculture
Animal and Plant Inspection Service
Riverdale, MD

This chapter describes how the USDA Animal and Plant Health Inspection Service (APHIS) guards against the importing of exotic plant pests and diseases, and how risk analysis is used in plant quarantine.

The 1994 North American Free Trade Agreement (NAFTA) and the 1995 General Agreement on Tariffs and Trade (GATT) were designed to reduce barriers to trade, investments, and services between the signing countries. NAFTA's Article 15 and GATT's Article 5 state that agricultural import decisions must be based on risk assessments that use scientific data and methods. We address these risk assessments first.

PEST RISK ANALYSIS

APHIS risk analysis has three components: risk assessment, risk management, and risk communication. Risk assessment addresses three questions: (1) What is the hazard (what is the undesirable event or outcome)? (2) How likely is the hazard to occur? and (3) What is the severity of impact of the hazard?

The APHIS approach to risk management also addresses three questions: (1) What can be done to prevent the hazard from occurring or to mitigate its consequences? (2) What are the options and associated costs, benefits, and tradeoffs? (3) What precedent will proposed actions to deal with this pest set for future actions?

Risk communication occurs throughout the process. Key communications must occur among public administrators, regulators, and the affected public to be sure that information and options are thoroughly considered. Communications also assure that affected parties understand the decision-making process and the actions that are taken.

INTERNATIONAL GUIDELINES

APHIS helped to draft standards for pest risk analysis through the North American Plant Protection Organization (NAPPO), along with other regional plant protection organizations and countries. The Food and Agricultural Organization (FAO) of the United Nations asked NAPPO to take the lead in drafting pest risk analysis guidelines for signers of FAO's International Plant Protection Convention (IPPC). In 1995, the FAO adopted international guidelines for pest risk analysis (1), and the United States was a major contributor to development of the guidelines. These guidelines are now used by APHIS as a basis for protecting the health of U.S. plant resources while facilitating trade.

Signers of the IPPC agreed to define a quarantined pest as "a pest of potential economic importance to the area endangered thereby and not yet present there, or present but not widely distributed and being officially controlled" (1). A pest analysis in the FAO guidelines includes risk assessment and risk management, but excludes risk communication. The purpose of the analysis is to determine whether a pest should be quarantined and how to manage risks that may exceed acceptable limits.

The analysis proceeds in three stages:

Stage 1. Initiation of the process. At the outset, one or more pathways for pest introduction or spread are identified, usually on a particular commodity. It then must be determined whether the pest qualifies for quarantine. If a previous risk analysis is not applicable, each pest must be evaluated to determine whether it meets requirements for quarantine.

Stage 2. Risk assessment. The stated purpose of risk assessment is to "determine whether a pest is a quarantine pest and to evaluate its introduction potential" (1). Estimates are made of potential geographic distribution, extent of actual distribution, official control measures already in place, and economic importance. Economic importance is based on potential for introduction and establishment as well as economic and environmental impact.

When we address the *potential for introduction*, we consider the likelihood of both entry and establishment. Factors that are considered include contamination of commodities or conveyances by the pest, survival of the pest in the shipping environment, ease or difficulty of detecting the pest during port-of-entry inspection, and frequency and quantity of pest movement into the area by natural means.

Factors that are considered in addressing *economic importance* are: (1) potential for establishment, based on availability, quantity, and distribution of hosts, environmental suitability, reproductive strategy of the pest, and method of pest survival; (2) spread potential after establishment, based on suitability of the natural or managed environment for natural spread of the pest, movement of the pest with commodities or conveyances, intended use of the commodity, and potential vectors of the pest; and (3) potential economic importance, based on the type of pest damage, crop loss, loss of export markets, increases in control costs, effects on ongoing integrated pest management programs, environmental damage, capacity to act as a vector for other pests, and perceived social costs such as unemployment.

Stage 3. Risk management. Pest risk management considers options that may include: (1) prohibiting entry of specific commodities from specific origins; (2) inclusion of a pest on a list of prohibited pests; (3) defining requirements to be satisfied before export, such as restricting exports to areas free from pests of concern, inspecting during the growing season, or inspecting and certifying shipments prior to export; (4) inspecting shipments at ports of entry; (5) treating the commodity at the time of or after entry; (6) requiring post-entry quarantine; and (7) requiring post-entry measures, such as restrictions on use of the commodity and control measures. The options are considered in terms of biological effectiveness, costs and benefits, impact on existing regulations, phytosanitary policy considerations, time to implement a new regulation, efficacy of the option against other quarantined pests, and commercial, social, and environmental impacts. Under GATT, the "minimal impact" principle must be followed: "Phytosanitary measures shall be consistent with the pest risk involved, and shall represent the least restrictive measures which result in the minimum impediment to the international movement of people, commodities and conveyances" (Article VI.2(f), IPPC).

APHIS RESOURCES AND ACTIVITIES

APHIS exists to protect the health of U.S. plant and animal resources. Its initial goal is to prevent entry of pests. When entry occurs, its purpose is to prevent spread. To accomplish its goals, APHIS makes preclearance and port-of-entry inspections and permit decisions. It applies quarantine treatments, conducts detection surveys, and attempts to eradicate exotic plant pests.

PORT-OF-ENTRY INSPECTIONS

Agricultural quarantine inspection is perhaps one of the most visible activities aimed at preventing the introduction of foreign pests and diseases. Seven days a week, some 1,300 USDA-APHIS inspectors are on duty at international airports, seaports, and border stations, to inspect passengers and baggage for plant and animal products that could be harboring pests or disease organisms. These APHIS Plant Protection and Quarantine (PPQ) inspectors check millions of passengers and their baggage each year for plant or animal pests and diseases that might harm U.S. agriculture. They also inspect ship cargoes, rail and truck freight, and mail from foreign countries. Table 1 provides selected inspection and interception data.

PPQ inspectors use a variety of techniques to augment the 75 x-ray units that check passenger baggage and mail for prohibited agricultural materials. Trained detector dogs sniff out prohibited fruit and meat. On leashes and under the constant supervision of their handlers, the USDA's "Beagle Brigade" has checked baggage from overseas for the past 10 years. Currently, APHIS has about 50 canine teams at

Table 1. Selected United States agricultural inspection data

	FY 91	FY 92	FY 93	FY 94	FY 95	FY 96
Ships inspected	52,119	53,374	47,887	53,270	55,205	52,974
Aircraft inspected	356,915	378,643	378,634	451,342	401,741	410,318
Passengers and crew inspected	53,999,534	58,103,711	56,920,156	62,548,979	65,645,734	66,119,960
Interception of plant material	1,527,922	1,723,004	1,474,569	1,442,214	1,583,687	1,567,886
Interception of pests	56,213	54,831	51,829	54,831	58,032	48,483
Interception of meat/ poultry products	205,407	246,878	224,340	281,230	223,392	264,001

21 U.S. airports. Dogs also are used at three post offices to inspect foreign mail. In addition to their actual function, the beagles serve as a symbol of the need to protect the nation's food supply from foreign pests. The Beagle Brigade was responsible for approximately 60,000 seizures of prohibited agriculture products in the 1994 fiscal year.

To report and monitor incipient populations of new introduced pests and domestic pests of significance to trade, PPQ works with the states in a project called the Cooperative Agricultural Pest Survey. Survey information on insects and plant diseases is entered into a nationwide database, the National Agricultural Pest Information System, found at the Internet site "http://www.ceris.purdue.edu/napis/." Information from this database can be accessed from anywhere in the United States.

PRECLEARANCE – CHECKING AT THE SOURCE

For this work, APHIS has experts stationed overseas. Often, it is most practical and effective to check and monitor commodities for pests or diseases at the source through preclearance programs. APHIS arrangements with other countries for preclearance programs are summarized in Table 2.

INTERNATIONAL PROGRAMS

Through direct overseas contacts, APHIS employees gather and exchange information on plant and animal health, work to strengthen national, regional, and international agricultural health organizations, and cooperate in international programs against certain pests and diseases that directly threaten U.S. agriculture. One example is the cooperative program for the Mediterranean fruit fly (*Ceratitis capitata*), commonly called the medfly in English or moscamed in Spanish. In 1977, the United States, Mexico, and Guatemala initiated a cooperative program known as the Moscamed Program to eradicate the medfly from Mexico and Guatemala, and to halt its northern spread. The Moscamed Program operates two facilities that produce and release sterile medflies: one in Metapa de Dominguez, Mexico, and one in El Pino, Guatemala. Mexico has been free of the medfly since 1982, except for outbreaks in southern Chiapas, adjacent to Guatemala.

Table 2. Examples of some USDA Animal and Plant Health
Inspection Service preclearance inspection programs

Country	Commodities
Argentina	Apples, pears
Australia	Apples, Nashi pears, pears, grapes
Belgium	Bulb inspection
Brazil	Mangoes (hot water treatment)
Chile	Stonefruit, berries, grapes, cut flowers, fruits, vegetables
Colombia	Mangoes (hot water treatment)
Costa Rica	Mangoes (hot water treatment), papaya
Ecuador	Mangoes (hot water treatment), melons (free zone)
France	Apples
Great Britain	Bulbs (inspection)
Guatemala	Mangoes (hot water treatment)
Haiti	Mangoes (hot water treatment)
Ireland	Bulb inspection
Israel	Bulb inspection
Jamaica	Thirty-one commodities including ugli fruit, cut flowers, papaya
Japan	Sand pears, Unshu oranges, Fuji apples
Korea	Sand pears, mandarin oranges
Mexico	Mangoes (hot water treatment), citrus (fumigate if not from free zone), apples, peaches

New Zealand	Apples, pears, Nashi pears
The Netherlands	Bulbs (inspection)
Nicaragua	Mangoes (hot water treatment)
Peru	Mangoes (hot water treatment)
Scotland	Bulbs (inspection)
South Africa	Apples, pears, plums, grapes
Spain	Lemons, clementines, Valencia oranges
Taiwan	Mangoes (hot water treatment)
Turkey	Bulbs (inspection)
Venezuela	Mangoes (hot water treatment)

COPING WITH INVASIONS

If, despite our best efforts, foreign pests or diseases slip past our border defenses, APHIS tries to control or eradicate them. Plant pest problems usually are handled by individual farmers, ranchers, and other property owners in cooperation with their state or local governments. However, when an insect, weed, or disease poses a particularly serious threat to a major crop, the nation's forests, or other plant resources, APHIS may join in the control work. Most pests and weeds that are targets of these programs are not native to the United States. When pests are new to this country, control techniques may not be available. PPQ may apply interstate quarantines and take other steps to prevent spread until effective control measures can be developed. PPQ has groups of specialists who deal with introductions of exotic plant pests. Known as "Rapid Response Teams", these groups have been mobilized on several occasions to combat costly infestations of Mediterranean fruit flies in California and Florida. Rapid Response Teams also responded when the destructive "A strain" of citrus canker was found in Florida orange groves and when karnal bunt, a fungal disease of wheat, was discovered in Arizona.

If a pest persists, and the program continues, other components are developed. For example, the medfly program includes survey, regulation, and control. In such a survey, APHIS works with state departments of agriculture to maintain trapping programs in areas susceptible to medfly establishment. When any medflies are collected in one of these high-risk areas, APHIS and state officials immediately implement a delimiting survey. Using the detection site as the focal point, field crews position additional traps to determine if an infestation exists and to locate and define the limits of the infested area. Regulations are imposed by federal and state quarantine officers, to prevent artificial within-state spread, and federal quarantine laws regulate the interstate movement of any article that may harbor the fly. State regulations control the movement of these articles going to uninfested areas. Articles regulated by state and federal authorities include all host fruits and vegetables present in the area. Open-air fruit and vegetable stands must provide protective covers for the produce to prevent infestation, and commercial and home-grown produce may not be moved without special inspection and treatment. For control, three kinds of treatments are used, alone or in combination, to eradicate the medfly: (1) aerial and ground application of bait spray; (2) the release of sterile male insects; and (3) application of insecticides to the soil under host trees.

IMPORT REGULATIONS

APHIS enforces regulations governing the import and export of plants and animals and certain agricultural products. Import requirements depend on both the product and the country of origin. Plants and plant materials usually must be accompanied by a phytosanitary certificate issued by a regulatory official of the exporting country.

The Plant Germplasm Inspection Station, a subunit of the National Plant Germplasm Quarantine Center, is a unique inspection station. Plant germplasm may be imported with a permit (with a few exceptions). Approximately 99% of propagative plant germplasm imported in small quantities for research purposes by the USDA Agricultural Research Service and its cooperators is inspected at the station. These importations of plant germplasm are treated if necessary, and then rejected or released to the importers. Before germplasm is exported to foreign countries, it is inspected at the station and certificates are issued if the conditions of entry can be met for the foreign country.

The seed examination facility (SEF) is responsible for enforcing the import requirements of the Federal Seed Act and the Federal Noxious Weed Act. PPQ officers at ports send seed samples to the SEF when they find suspected noxious weed seeds. Under the Federal Seed Act, seed shipments may be refused for incorrect or incomplete labeling or for excessive noxious weed content. The SEF coordinates the post-entry quarantine program, which is a federal-state cooperative program. PPQ issues the permits to release the materials. The states conduct the inspections. Hawaii and Puerto Rico have their own post-entry quarantine programs, and APHIS supplies them with tags, ties, and the necessary forms. Seventy PPQ liaison officers have been trained in post-entry quarantine to assist with the program. They are located throughout the four PPQ national regions.

To import live plant pests for research or as biological control agents, a permit is required. Upon receipt of shipments of live plant pests, the packages are inspected and forwarded to the recipient if the shipment meets U.S. entry requirements. Detailed requirements and applications for a permit are available at the Internet site at "http://www.aphis.usda.gov/ppq/permits."

APHIS cooperates with the U.S. Department of the Interior in carrying out provisions of the Endangered Species Act that deal with imports and exports of endangered plants, animals, or birds. APHIS inspectors at ports of entry are trained to identify these species and notify the Department of the Interior of any species found during inspections that are protected under the Convention on International Trade in Endangered Species.

EXPORT REGULATIONS

To facilitate agricultural exports, APHIS officials certify the health of plants and animals that are shipped to foreign countries. PPQ provides assurance that U.S. plants and plant products meet the plant quarantine import requirements of foreign countries. This assurance is in the form of a phytosanitary certificate, issued by PPQ or its state cooperators. During FY 1994, 271,000 phytosanitary certificates were issued for exports of plants and plant products worth $23 billion.

In addition to certifying the health of agricultural exports, APHIS officials mount a proactive approach to the marketing of U.S. crops and livestock overseas. For instance, APHIS and Foreign Agricultural Service officials helped maintain U.S. wheat exports after the March 1996 discovery of an outbreak of karnal bunt, a fungal disease of wheat, in Arizona. Their efforts helped the United States maintain its status as the world's leading wheat exporter ($5.5 billion in 1995).

TRAVELERS ENTERING THE UNITED STATES

Many passengers entering the United States do not realize that one piece of fruit packed in a suitcase has the potential to cause millions of dollars in damage to U.S. agriculture. Prohibited fruits and vegetables can carry a whole range of plant diseases and pests.

United States travelers planning a trip abroad should check the list of approved products, which is not all-inclusive. Even though an item brought into the U.S. may appear on the list of approved products, travelers are still responsible for declaring to a federal inspection officer every agricultural product in their possession. Travelers leaving the United States with any agricultural products may not be allowed to bring them back when they return. A U.S. border official should be consulted before travelers take such goods across the border. The APHIS Internet web page (http://www.aphis.usda.gov) provides up-to-date information on this and other related topics.

FRUITS, VEGETABLES, AND PLANTS

Some fruits, vegetables, and plants may be brought into the U.S. without advance permission, provided they are declared, inspected, and found free of pests. However, a permit must be obtained in advance to bring in certain plants and plant parts intended for growing. If a particular plant is not allowed entry, then a risk analysis may need to be conducted and regulations governing importation of this plant amended. To bring back threatened plant species, a permit or certificate from the country of origin, as well as from the U.S. Department of the Interior's Fish and Wildlife Service, is usually required.

OTHER BIOLOGICAL MATERIALS

A permit is required to bring in most organisms, cells and cultures, monoclonal antibodies, vaccines, and related substances, whether of plant or animal origin. This category includes organisms and products used in the biotechnology industry.

SOIL, SAND, MINERALS, AND SHELLS

Soil-borne organisms threaten both plants and animals. If travelers visited a farm or ranch overseas, agricultural inspectors may have to disinfect their shoes or clothes. Vehicles must also be cleaned of soil. Soil, earth, or sand is not allowed, although an ounce or less of decorative beach sand is allowed. Rocks, minerals, and shells are allowed, but all sand and soil must be removed. Products grown in soil (like shamrocks and truffles) must be free of soil.

Literature Cited

1.   Food and Agriculture Organization. 1995. Guidelines for Pest Risk Analysis. 13 pages.

2.   USDA-APHIS-PPQ. 1996. Delivery of Plant Protection Programs in the United States – The Role of the Animal and Plant Health Inspection Service. 4 pages.

Assessing Exotic Threats
to Forest Resources

William E. Wallner
USDA Forest Service
Northeastern Center for Forest Health Research
Hamden, CT

Over the past 150 years, North American forests have proved
highly vulnerable to imported pests, and increased U.S. imports of
forest products suggest major threats now and in the future. Ways
must be found to protect our forests as we import the products we
need, but our present defenses are far from reassuring.

Forests are particularly vulnerable to new pests because the control
measures used in agriculture are often logistically, economically, and
ecologically impractical (23). In addition to being expensive,
treatments to eradicate or control exotic pests are often controversial
(1,19). Certainly, therefore, exclusion is the preferred tactic.

In this article, I show how threats to forests from invading exotic
pests are being assessed. I also show that use of assessment results
should not be the same for forests as for agricultural plants.

INCREASED IMPORTS MEAN INCREASED RISKS

In recent years, the volume of timber harvested on public land
in the United States has decreased sharply. Manufacturers of wood
products in the United States, therefore, have increased their imports
of unprocessed wood. Other developed nations also have increased
their imports, and unfavorable consequences have been predicted
(12). Other major pathways for movement of exotic pests include
ornamental trees, wood packing materials, and germplasm for
research and development of improved trees.

Consider *Eucalyptus*, a genus widely planted around the world
for wood fiber as well as ornamental specimens. While a number of
insects and diseases attack some of the 700 species of this genus, few
cause critical damage. One insect that does is the Eucalyptus weevil.
Both adults and larvae are host-specific defoliators of this genus.
Attacks were reported on introduced *Eucalyptus* species in New
Zealand in 1890, in South Africa in 1916, in Argentina in 1925, in
Mauritius in 1935, in Madagascar in 1950, in some Mediterranean

countries of Europe in 1975, in Spain and Brazil in 1992, in California in 1994, and in Chile in 1998. Clearly, this pest has demonstrated an ability to move around in the world. Since *Eucalyptus* species are so widely used as ornamentals and trees for forest plantations, many individual countries and trading blocs have import regulations to prevent the entry of this pest.

Fortunately, many manufactured wood products, such as paper, plywood, and kiln-dried lumber have little risk of carrying insects or diseases. These products are essentially sterilized during manufacture. That is not true for logs, wood chips that have not been fumigated, and live plant materials.

RISK ANALYSIS FOR FOREST ECOSYSTEMS

The International Plant Protection Convention (IPPC) was adopted by 79 nations in 1951 to prevent the spread of dangerous plant insects and diseases (2). The IPPC adopted 16 plant-quarantine principles that reflect the importance of a technical basis in relevant international trade decisions (3). While most of these mechanisms are appropriate for forestry, the recommendations for risk analysis and some other measures are not as easily applied in forestry as in agriculture.

Quarantine is fundamental to preventing the entry of exotic pests. Quarantine regulations rely on risk analysis to set standards (5), but the standards for forest products have been the subject of much debate and litigation. Theoretically, the ecological risk associated with an introduced organism can be estimated by cost-benefit analysis, impact analysis, or pest risk analysis (PRA). The PRA is preferred, because it incorporates a number of important ecological, economic, and socio-political factors. Regardless of the approach that is followed, however, the person doing the analysis must be able to answer two questions: (1) what is the probability that the introduced species will be harmful? and (2) how harmful is the introduced species likely to be?

Because forest ecosystems are highly complex, and most forest pests are not thoroughly understood, the answers to these key questions often represent little more than speculation.

The behavior of an organism that is introduced into a new and complex environment is next to impossible to predict. Certainly, if the host plant genus exists in both the native country and the potential receiving country, the pest should be considered a potential threat. Organisms that are relatively benign in their native habitat, however, can become major pests when introduced into a new environment that lacks their native parasites and predators.

For example, in 1988, two seed buyers from China collected slash pine cuttings from a south Georgia seed orchard to graft into a seed orchard in Guangdong Province, China. They did not notice a tiny mealybug (*Oracella acuta*) hidden beneath needle scales (Fig. 1). The cuttings were propagated, and the improved pine genotype, along with the hitchhiking mealybugs, was planted on 12.4 hectares. After two years, the Chinese noticed a heavy growth of sooty mold, a black fungus that grows on honeydew from sucking insects. The fungus alerted them to the presence of the mealybug, which had been imported without any of the natural enemies that keep it in check in the southern United States. By 1995, the pest had spread unchecked

Fig 1. The pine mealybug was accidentally introduced into China from the United States on pine cuttings, with disastrous results. Photo courtesy of G.L. DeBarr, USDA Forest Service, Southern Research Station.

Table 1. Cost of preparing U.S. pest risk assessments for wood products*

Country of origin	Year	Cost
Siberia	1991	$500,000
New Zealand	1992	$ 28,000
Chile	1993	$ 51,000
Mexico	1996	$ 50,000
South America	1998	$ 60,000

*Does not include salaries of federal scientists in preparing these documents.

over 200,000 hectares and was reducing pine growth by as much as 30% (13). This outbreak could not have been anticipated based on the behavior of the mealybug in its homeland, where natural enemies keep populations low.

The most direct way to address the uncertainty of pest introductions is a process that analyzes potential introductions based on a specific commodity pathway (9). This approach, which has been used to assess pest risk to U.S. forests from the import of unprocessed logs, is expensive (Table 1). The large disparity in costs between the PRA for logs from Siberia and other countries was principally due to outside contracting costs. However, the Siberian PRA also established the procedures and standards for three subsequent PRAs. These detailed economic analyses of organisms of concern served as the basis for quarantine regulations and mitigation considerations for the importation of logs.

PATHWAY ASSESSMENTS

Every region of the world has its complement of exotic pests, but the New World has received a disproportionate share (8). Some of the more damaging pest species are listed in Tables 2 and 3, along with their pathways of introduction.

Table 2. Forest insect pests introduced into the United States

Pest	Origin	Introduction pathway
Gypsy moth (European)	France	Scientific investigations
Gypsy moth (Asian)	Russian Far East	Ships, containers
Balsam woolly adelgid	Europe	Nursery stock
Larch casebearer	Europe	Nursery stock
European elm bark beetle	Europe	Raw veneer logs
Red pine scale	Asia	Nursery stock
Hemlock woolly adelgid	Asia	Nursery stock?
Pine shoot beetle	Eurasia	Ship dunnage?
Spruce beetle	Eurasia	Ship dunnage
Beech scale	Europe	Nursery stock
Asian long-horned beetle	Asia	Wood packaging

In estimating the threat of an exotic organism to forest ecosystems, pathway analysis is a useful, but imperfect tool. Pathway-initiated PRAs have been used successfully for animal diseases and agricultural commodities. Quantitative assessments for forestry rarely have been attempted because of the complexity of forest resources and the difficulties associated with assigning numerical values to a multitude of pests and non-commodity values. Instead, PRA elements are

Table 3. Forest diseases, their origin, and pathway of introduction into the United States

Pest	Origin	Introduction pathway
Chestnut blight	Asia	Nursery stock
Dutch elm disease	Europe	Raw veneer logs
White pine blister rust	Eurasia	Nursery stock
Larch/poplar Melampsora rust	Eurasia	???
Port Orford cedar root disease	Asia	Nursery stock?
European larch canker	Europe	???
Beech bark disease	Europe	Nursery stock?

assigned a risk value along with a certainty code (9). This procedure relies on the experience and knowledge of the assessor as well as available biological information.

For example, for assessing pest risk associated with unprocessed wood, literature surveys and the opinions of international entomologists, pathologists, and nematologists determine an initial list of potential pests.

From this list, individual PRAs are conducted for organisms found in specific pathways, such as on or under the bark of logs. A generic pathway evaluation has the advantage of anticipating the suite of organisms found in that pathway. In this approach, mitigation measures can be prescribed for unknown as well as known pests that may use the pathway.

Pathway analysis is not always conclusive. Some organisms are not necessarily affiliated with forestry products, such as logs, lumber, chips, and nursery stock. Such non-product pathways are difficult to predict. As global trade increasingly homogenizes the earth's biota, new pests and unique pathways are likely to emerge. For example, China recently has become a major U.S. trading partner, greatly escalating the probability of pest importation. Wood in spools for wire, rope, packing material, pallets, and bracing for transporting cargo in containers, have carried exotic pests. Via this pathway the Asian long-horned beetle, *Anoplophora glabripennis* (Fig. 2), gained entry into a number of North American locations and resulted in major infestations in Chicago and New York. Native to China, Japan, and Korea, the longhorned beetle attacks and kills numerous hardwoods (poplars, willows, and fruit trees) by boring into the trunks and branches. In the U.S., streetside maples and horsechestnuts have been infested (18). Eradication efforts that include tree removal, chipping, and burning have been controversial issues in these urban environments. The most valuable commercial resource at risk is the sugar maple products industry because sugar maple is a highly preferred host for this insect. Also at risk are fall foliage and tourism

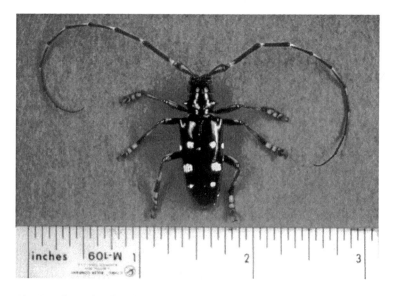

Fig. 2. The Asian longhorned beetle, *Anoplophora glabripennis*, has been introduced into North America at several locations in association with wood packing material.

Fig. 3. Discarded wood used for packing and shipping cargo. Such wood usually is of low quality, is seldom kiln-dried, often contains bark, and sometimes contains exotic pests.

industries of the northeastern United States and Canada. Introductions of Asian long-horned beetles via wood packing materials have led the USDA Animal and Plant Health Inspection Service to enact regulations to close this pathway. However, because wood packing material is used extensively internationally (Fig. 3), compliance with such regulatory efforts has serious trade implications.

QUARANTINES

Quarantine of an organism often is based on its performance in another country. Yet, such information often leads to underestimation of the risk. Before any risk assessment can be undertaken, the most important potential quarantine pests must be listed. The European Plant Protection Organization (EPPO) has begun to compile such a list. To date, this list has relatively few potential forest pests.

In the United States, a targeted quarantine pest must meet one of the following criteria:

1. Nonindigenous; not present in the United States (e.g., nun moth, *Lymantria monacha*) (15);

2. Nonindigenous; present in the United States but capable of expanding its range (e.g., Asian longhorned beetle, *Anaplophora glabripennis*); or

3. Nonindigenous or indigenous; genetically different or able to vector other nonindigenous species (e.g., Asian biotype of gypsy moth, *Lymantria dispar*) (20).

Many difficulties plague those who conduct PRAs. There is no clear consensus on how genetic variants of the same species should be considered in PRAs, and genetic variability in behavior and successful establishment is difficult to ascertain. Dirty/clean pest lists lack accuracy. Pest targeting has limitations. Pathway analyses for forest pests cannot always predict hazards. Cost-benefit analyses are difficult to complete, and economical and environmental impacts are estimated only with wide variation. Despite these difficulties, the damage potential to natural forest ecosystems by an introduced organism requires that we DO SOMETHING! However, we must proceed with caution and scrutiny.

EXOTIC FOREST PEST LIST

The North American Forest Commission of the United Nations Food and Agriculture Organization has chartered its Insect and Disease Study Group to compile a list of exotic insects and pathogens with potential to cause significant damage to forest resources in Canada, the United States, and Mexico. The list will be annotated with specific information used in the rating process. It will incorporate known pest pathways (14,15,16,17). Another list is being compiled of the highest priority pests currently regulated in North America or proposed for testing by international experts. Detailed pest analyses will be undertaken and a descriptive leaflet will be developed for each pest. Access via the Internet will encourage participation and these databases will contain information for regulatory research and control purposes (4). The website for the North American Exotic Forest Pest Information System is: www.exoticforestpests.org. Other countries are developing similar lists.

DISCUSSION

Given the high uncertainty about effects of imported pests, some people might question the value of making risk assessments for forest pests. I argue for them because they provide a useful way to organize the information we have. The difficulties with risk assessments are most likely to be caused by faulty interpretation. Several key questions need to be addressed during interpretation: (1) How is uncertainty addressed? (2) What conclusions do we reach when the assessment shows a small risk of a catastrophic event? And (3) what steps do we take to close pathways that have already been used by damaging pests in the past?

Experience shows that the pestilence of an organism cannot be predicted from its status in its native country. For example, only 18% of immigrant insects and mites in the United States behaved exactly as one would have expected from their behavior in their country of origin (7). Island nations, such as Australia and New Zealand, are particularly vulnerable to invading pests. In managing risks, they

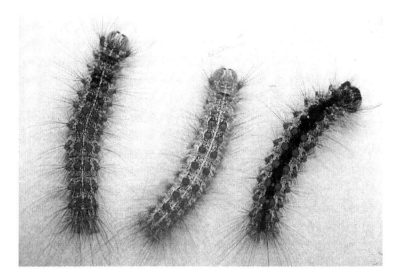

Fig. 4. Asian gypsy moth, *Lymantria dispar*, larvae reflecting variation in patterns of color not found in the strain of gypsy moth introduced from Europe. The Asian strain is more dangerous because larvae feed on a wide variety of trees, including conifers, and, unlike its European cousin, females can fly.

favor an emphasis on conservatism in the absence of adequate risk data (23). Ideally, biological and socio-economic data should be used to measure the level of risk, but such data are lacking for most complex forest ecosystems.

Even a cursory examination of entry pathways for important forest pests shows that nursery stock, unprocessed logs, and wood packing materials have been a big problem. Ways must be found to close these pathways, even if the specific pest that may use the pathway is not identified. As we import goods from less developed countries, there is a good chance of importing a pest that has not yet been described and named, much less rated for damage potential.

If old pathways are closed, new ones will create surprises for us. Russian grain ships and U.S. military equipment have carried the Asian gypsy moth (Fig. 4) into the United States and Canada. Unlike the European strain already present in the U.S., Asian gypsy moth females are active flyers. They are readily attracted to artificial lighting, where they lay masses of 500 to 1000 eggs. Russian grain ships carried the larvae to West Coast ports, where they became established. Costs of eradication programs over a four-year period were approximately $17 million, and some of the people living near the ports objected to the use of the bacterium *Bacillus thuringiensis* for control (Fig. 5).

Fig. 5. Aerial application of *Bacillus thuringiensis* sprays to control Asian gypsy moth, *Lymantria dispar*, in Vancouver, British Columbia.

While awaiting shipment to the United States from Germany, some military equipment became infested with Asian gypsy moths. The passage from Europe to the United States takes only 10 days, and the insects easily survived it and established a reproducing population at a military terminal in North Carolina. The cost of a three-year eradication program was $6 million.

During this period, military containers and vehicles were inspected and shipped from North Carolina to 48 locations throughout the United States. If the vehicles or containers had been infested, the Asian gypsy moth could have been spread over much of the nation (21).

Four recent PRAs (14,15,16,17) predict cumulative losses of $60 billion to potential pests from individual countries. While these estimates may seem excessive, they are consistent with other estimates (10). When we honestly arrive at such assessments, we must not be afraid to report such large numbers. Big numbers capture the interest of decision-makers. Small numbers normally do not.

Literature Cited

1.   Bonford, M., and O'Brien, P. 1995. Eradication or control for vertebrate pests? Wildl. Soc. Bull. 23:249-255.

2.   Food and Agriculture Organization. 1951. International plant protection convention. Food Agric. Org., Rome, Italy.

3.   Food and Agriculture Organization. 1995. International standards for phytosanitary measures. Reference standard. Principles of plant quarantine as related to international trade. ISPM Publ. No. 1. Food Agric. Org., Rome, Italy.

4.   Lorimer, N. 1997. Draft implementation plan: Exotic forest pest list project of North American Forestry Commission, Insect and Disease Study Group.

5.   Mathys, G., and Baker, E.A. 1980. An appraisal of the effectiveness of quarantines. Annu. Rev. Phytopathol. 18:85-101.

6.   Mattson, W.J., Niemela, P., Milleers, I., and Inguanzo, Y. 1994. Immigrant Phytophagous Insects on Woody Plants in

the United States and Canada: an annotated list. USDA Forest Service GTR-NC-169.

7.   McGregor, R. C. 1973. The emigrant pests. Report to administration. USDA Animal and Plant Health Insp. Serv., Hyattsville, MD.

8.   Niemela, P., and Mattson, W.J. 1996. Invasion of North American forests by European phytophagous insects. BioScience 46:741-753.

9.   Orr, L., Cohen, S.D., and Griffin, R.L. 1993. Generic non-indigenous pest risk assessment process. USDA Anim. and Plant Health Insp. Serv., Washington, D.C.

10.   Office of Technology Assessment. 1993. Harmful non-indigenous species in the United States. Publ. OTA-F-65. Off. Technol. Assess., Washington, D.C.

11.   Sailer, R. I. 1983. History of insect introduction. Pages 15-38 *in:* Exotic plant pests and North American agriculture. L. Wilson and C.L. Graham, eds. Academic Press, New York.

12.   Sedjo, R. A. 1993. Global consequences of U.S. environmental policies. J. For. 91(4):19-21.

13.   Sun, J. DeBarr, G.L., Berisford, C.W., and Clarke, J.R. 1996. An unwelcome guest in China: A pine-feeding mealybug. J. For. 94:27-32.

14.   Tkacz, B. M., Burdsall, H.H., DeNitto, G.A., Eglitis, A., Hanson, J.T., Kliejunas, J., O'Brien, J.G., Smith, E.L., and Wallner, W.E. 1998. Pest risk assessment of the importation into the United States of unprocessed *Pinus* and *Abies* logs from Mexico. USDA Forest Service Misc. Publ. FPL GTR 103.

15.   U.S. Department of Agriculture, Forest Service. 1991. Pest risk assessment of the importation of larch from Siberia and the Russian Far East. USDA For. Serv. Misc. Publ. 1495.

16. U.S. Department of Agriculture, Forest Service. 1992. Pest risk assessment of the importation of *Pinus radiata* and Douglas-fir logs from New Zealand. USDA For. Serv. Misc. Publ. 1508.

17. U.S. Department of Agriculture, Forest Service. 1993. Pest risk assessment of the importation of *Pinus radiata, Nothofagus dombeyi,* and *Laurelia philipiana* logs from Chile. USDA For. Serv. Misc. Publ. 1517.

18. U.S. Department of Agriculture, Forest Service. 1997. Asian Cerambycid Beetle, A New Introduction. Pest Alert NA-PR-01-97.

19. Usher, M. B. 1989. Ecological effects of controlling invasive terrestrial vertebrates. Pages 463-489 *in*: Drake, J.A., Mooney, H.A., diCastri, F., Groves, H.R., Kruger, F.J., Rejmanek, M., and Williamson, M., eds. Biological invasions: a global perspective. SCOPE Ser. 37. John Wiley and Sons, Chichester, UK.

20. Wallner, W. E. 1996. Invasive pests (biological pollutants) and U.S. forests: whose problem, who pays? EPPO Bull. 26:167-180.

21. Wallner, W.E. 1997. Global Gypsy – The Moth That Gets Around. Pp. 63-76 *in:* K.O. Britton, ed. Proceedings, Exotic Pests of Eastern Forests. Nashville, TN.

22. Wallner, W.E., Humble, L.M., Levin, R.E., Baranchikov, Y.N., and Carde, R.C. 1995. Response of adult lymantriid moths to illumination devices in the Russian Far East. Jour. Econ. Entomol. 88:337-342.

23. Wylie, F. R. 1989. Recent trends in plant quarantine policy in Australia and New Zealand and their implications for forestry. N. Z. J. For. Sci. 19:308-317.

Political and Economic Barriers
to Scientifically Based Decisions

Faith Thompson Campbell
Western Ancient Forest Campaign
Springfield, VA

Some new invaders will arrive despite our best endeavors to exclude them. Others will enter due to carelessness – failure by responsible parties to place adequate priority on preventing introduction of additional harmful species and failure to utilize all available methods to minimize such introductions.

We face the inevitability of new introductions facilitated by the rapid expansion of global commerce. In its recent evaluation of inspection efforts by the USDA Animal and Plant Health Inspection Service (APHIS), the General Accounting Office (GAO) noted "[s]everal developments . . . challenging APHIS' ability to effectively manage its inspection program. Key among these is the rapid growth in international trade and travel. . . ."

In addition to stopping new arrivals, we must halt the continued spread of the several hundred harmful alien species already widely disseminated in the country. We must closely monitor potential invaders that have already been introduced and move promptly against them when they give the first indication of an ability to spread. For example, according to John Kartesz of the North Carolina Botanical Garden, some 4,000 foreign plant species have been documented as having spread outside cultivation in this country. Thousands more are in our gardens. Some proportion of these probably will invade relatively undisturbed habitats. If only 10% of the 8,000 species in cultivation in Hawaii prove invasive, the number of invasive plant species would double, along with the workload of those islands' already overwhelmed resource managers (2).

Another gap in our defenses is deliberate import. Unwise decisions provide a particularly important "pathway" for invasive plant species. A surprisingly high proportion (62%) of the plant species now devastating our native ecosystems are still offered for sale (Campbell, unpublished). New plant species continue to be imported without being screened for "weediness".

OUR GOVERNMENT'S POLICIES ARE INADEQUATE
TO PROTECT OUR ECOSYSTEMS

Underlying all other problems plaguing our pest exclusion programs are hostility toward governmental regulation and the current emphasis on the free market. Three decades ago, Garrett Hardin pointed out that markets do not perform well as protectors of common resources (1). Individuals profit from importing and disseminating products, such as a new ornamental plant, while the environment and society generally absorb the cost of combating any species that proves to be invasive. There is a role for voluntary actions by commercial entities and the public, but protecting ourselves and our environment from these invasions also calls for regulation. The federal government is the entity charged by the Constitution with regulating imports. We need tougher regulations, and better enforcement.

Another problem is our government's excessive focus on trade and customer service – policies that undercut effective scrutiny of cargoes to detect hitchhiking invaders. Again, the GAO notes that APHIS' adoption of a trade-facilitation policy poses a "challenge" to the same agency's carrying out its task of "minimizing the risks of infestation and disease. . . ." APHIS alone is responsible for excluding pests, while several other agencies promote international trade (5).

APHIS' enforcement priorities also seem inconsistent with the real risk. I believe there is less risk from travelers carrying fruit and other food items for personal consumption, than from commercial shipments of a wide range of commodities, such as logs (3,4) and nursery stock. Preventing additional introductions of pests hitchhiking on imported nursery stock will require more intensive inspections and, probably, new technologies. The GAO has criticized APHIS for not inspecting cargoes with sufficient thoroughness (5).

APHIS has adequate authority to inspect shipments and, if necessary, to quarantine or deny entry to shipments harboring known or suspected pests. Yet, at least five new exotic pests have entered the country since 1991. Why? I attribute the introductions to the inherent difficulty of the quarantine task, shortcuts in inspections forced by failure of resources to keep pace with expanding workload (5), plus – in my view – APHIS' failure to appreciate the value of non-agricultural resources put at risk and thus to devote resources to protect them.

Statutes on introduction of potentially invasive plant species must be strengthened significantly. We must require that new introductions be screened for "weediness". We should also regulate imports of a much higher proportion of exotic plant species already known to be invasive. The Federal Noxious Weed Act of 1974 has done little to curb importation of most foreign plant species that are known to be invasive. The other gaps in this Act and its interpretation to date undermine its effectiveness in protecting our ecosystems from invasive alien plant species. Chief among these gaps are insufficient priorities given to species that invade primarily natural environments, lack of authority for APHIS to list widespread species, confusion as to whether APHIS has authority to outlaw interstate shipments of noxious weeds, and lack of emergency authority to restrict imports of harmful plant species before they have been officially designated as "noxious".

While a few members of Congress have sought recently to strengthen the Federal Noxious Weed Act, none is willing to sponsor a bill that would require pre-import screening for weediness. If resource managers are to succeed against exotic pests or weeds, they need a respite from new invaders. The current level of protection, already inadequate, is likely to be further relaxed with the new emphasis on reducing "non-tariff barriers to trade" associated with adoption of the World Trade Agreement. Revisions to the International Plant Protection Convention (IPPC), adopted in November 1997, do not reflect the growing scientific concern about the impacts of invasive alien species and will not allow governments to exclude potential invaders in a timely manner.

The new IPPC restricts a country's right to exclude alien pests or weeds. If a particular pest or weed species is already "widespread" in the United States, APHIS may not impose a quarantine to prevent importation of additional individuals of that species. Even if the species is not "widespread", APHIS may institute a quarantine only if here is an "official" control program targeted on that species. Additional introductions made virtually certain by this provision will add to the numbers of the pests in the country, expand the geographic areas into which pests have been introduced, and increase the genetic variability within the pest population – thereby enabling the pest to spread to new regions or with greater virulence. Kudzu is a prime example of a known invasive weed that does not meet these standards (Fig. 1). APHIS cannot prevent the import or planting of kudzu because it is already widespread.

As for potentially invasive species not yet in the country, the right to take protective action depends on regulatory agencies promptly predicting whether that insect, fungus, or plant species will invade and cause economically measurable damage. The natural environments

Fig. 1. Kudzu is a noxious weed invading up to 120,000 acres per year in the southeastern United States.

and agricultures of importing countries will bear the brunt of any mistakes.

I believe that for the IPPC standards to be fully "science based", they should allow countries to treat an exotic species as "guilty until proven innocent" and require that its invasive potential be evaluated before its importation is permitted. While the "guilty until proven innocent" approach represents a complete reversal from current practice in most countries, it is the approach most recommended by ecologists. Already, New Zealand and Australia are moving in this direction.

OBSTACLES

If we are to tackle the scientific problems that hinder effective pest exclusion programs, we first have to spark the political will to address the very concept of damaging invasive alien species. We must compete for people's attention against a wide range of other enticements. Then we must convince people that the harm caused by invasive species impacts them directly. That task requires more complete information and the help of marketing experts to convey information in ways that will motivate people to act. For our message to be absorbed, we must repeat it frequently – as does that great American success story in motivation, the advertising industry.

Experts on biological invasions should aggressively compile information on impacts and effective control methodologies and press their colleagues in the environmental education profession to incorporate that information into their materials. The general public needs to be informed.

Some of the missing information is at a very basic level. Most Americans recognize only a few plant species. They have no idea which ones are native and which are exotic. How do we help our neighbors learn to recognize and appreciate our own flora, and, in reverse, to see the rapid proliferation of alien plants? This education effort is especially important in forests, where exotic shrubs, trees, and herbaceous flora replace native counterparts. With regard to forest pests, we need to find ways to keep people aware of the species like the American chestnut that have virtually disappeared, so that they understand the dimensions of the problem.

We also need to fill more traditionally recognized information gaps, such as hard data on the extent of exotic plant invasions and the ecological damage caused by the various alien species. Whenever possible, we should attach some economic loss figures to the damage.

Building an effective program to contain and suppress plant invasions will require the active cooperation of the horticultural trade. People who earn their living by selling plants and who feel competitive pressures to find "new and improved" varieties to offer, will need to consider the greater good and to forego sales of certain species. Furthermore, increasingly the decisions will be based on a prediction that a species may be invasive, rather than on observed facts. We are asking the industry to accept significant – but I think necessary – sacrifices.

Finally, we will have to persuade people to support significant increases in spending and regulatory authority by federal agencies.

We also need greater support from the mainstream environmental organizations. Here we face another challenge. The "weed" issue scrambles their traditional alliances. Environmentalists will be asked to join forces with chemical companies and livestock ranchers and support active management of areas they have thought of as "pristine". Wildlife organizations will have to abandon their own past advice regarding plantings to "enhance" habitat, fishing enthusiasts will have to stop stocking introduced fish into lakes and streams, and people concerned about humane treatment of animals will have to accept the killing of various exotic mammals and birds.

A fundamental problem remains: the threat from invasive alien species is often seen as fundamentally different from other environmental threats, and therefore needing a completely new type of response. I argue that invasive aliens are not fundamentally different. That is why I welcome use of the term "biological pollution."

No matter whether the threat to our biodiversity stems from chemical pollutants, overharvesting, suburban sprawl, draining of wetlands, or biological pollutants, we cannot put the environment back the way it was. We can reverse the effects or restore some areas and minimize future damage. I think our message should emphasize the similarities with other environmental threats, even as we acknowledge the need for different actions to address this specific manifestation.

Literature Cited

1.   Hardin, G. 1968. Tragedy of the Commons. Science. Vol. 162. Pp. 1243-48.

2.   Smith, Dr. Clifford. University of Hawaii. Eastern Weed Summit, November, 1995.

3.   United States Department of Agriculture, Forest Service. 1992. Pest Risk Assessment of the Importation of *Pinus radiata* and Douglas-fir Logs from New Zealand. Miscellaneous Publication No. 1508, October, 1992.

4.   United States Department of Agriculture, Forest Service. 1993. Pest Risk Assessment of the Importation of *Pinus radiata, Nothofagus dombeyi,* and *Laurelia philippiana* Logs from Chile. Miscellaneous Publication No. 1517, September 1993.

5.   United States General Accounting Office. 1997. Agricultural Inspection: Improvements Needed to Minimize Threat of Foreign Pests in Diseases. Washington, D.C. May 1997. GAO/RCED-97-102.

Fighting Back!

Kerry O. Britton
USDA Forest Service
Southern Research Station
Athens, Georgia*

We now know that biological pollution threatens the complex species structure and ecological function of our forests, our grasslands, and our lakes and streams, as well as our crops. What can be done to prevent more introductions, further spread, and increasing losses?

We cannot afford to do nothing. The U.S. Congress Office of Technology Assessment (OTA) report, *Harmful Non-Indigenous Species In The United States,* estimates $97 billion in losses from just 79 exotic species from 1906 to 1991 (13). This is probably only a fraction of the true cost of the 4,500 exotic species present in the United States (see chapter by Windle, this volume).

Invasive exotics often lull us into a false sense of security by what is known as a "lag phase". This period of slow growth can last 80 to 100 years, during which time the exotic may be adapting to its new environment, or just building up "critical mass" for sudden expansion. Thus, we may perceive the exotic organism as "harmless", or even "beneficial", until it suddenly spreads explosively (12).

RECOGNIZE THE PROBLEM AND **KEEP** IT OUT

The first step toward solving any problem is recognition. Please share this book with people who care about natural resources. Most people are unaware of the magnitude of the threat biological pollutants pose to both agricultural and native ecosystems. Past ignorance is the source of biological pollution.

No other pollutant is smuggled into the home country in the suitcases of well-meaning but misguided tourists enchanted by exotic species. No other pollutant is introduced with fanfare and proudly

*Present address: USDA Forest Service, Forest Health Protection, Arlington, VA

propagated by horticulturists, or by land managers working to control erosion, or to benefit a particular group of wildlife, without recognizing the deleterious impacts . . . until it is too late. Pest exclusion is the cheapest way to prevent biological pollution.

Sometimes exotic nurseries stock the vehicle of destruction, rather than its source. Chestnut blight is a classic example. The fungus probably arrived in this country on Chinese chestnut trees (5). Since Chinese chestnut has genetic resistance to the fungus, infected trees would appear healthy. Many other exotic plant diseases, e.g., dogwood anthracnose, pitch canker, white pine blister rust, and poplar leaf rust, were also introduced through movement of nursery stock.

Today, imported nursery stock is highly scrutinized, particularly plants in soil, but movement of plant parts not growing in soil is much less restricted. Offshore propagation of cuttings is becoming commonplace, yet also bears hazards.

RECOGNIZE THE PROBLEM AND **STAMP** IT OUT

Gypsy moth (*Lymantria dispar* L.) is another infamous scourge, responsible for the devastation of forests from Maine to Michigan and Virginia. To date, billions of tax dollars have been spent studying gypsy moth and spraying control agents to eliminate new infestations and "slow the spread". How much less expensive would it have been if entomologists, who were notified by E. Leopold Trouvelot – the French silkworm enthusiast who accidentally released gypsy moth in his Boston suburb around 1869 – had acted immediately to eradicate the initial infestation? (Fig. 1).

It is estimated that 4,600 acres PER DAY of land in the United States are being invaded by exotic weeds (see chapter by Campbell, this volume). About half of this area is public land – land that our elected officials have decided for one reason or another should be protected for posterity by federal or state ownership. Yet, public land managers are fighting a losing battle against exotic weeds. They cannot win it alone.

Many plants imported as desirable species have been found later to be extremely invasive weeds. Kudzu, imported as an ornamental in 1876, and widely planted for erosion control in the 1930s, has been estimated to cover 7,000,000 acres. This pernicious vine has steadily encroached into farmers' fields and engulfed forest trees in the South, and now has gained footholds from Connecticut to east Texas. *Melaleuca,* Japanese and Chinese privets, multiflora rose,

Fig. 1. Men in trees: Initial efforts to eradicate gypsy moth in New England involved climbing trees and destroying egg masses. Photo courtesy of Bill Wallner.

Japanese honeysuckle (Fig. 2), and Chinese wisteria are other examples of "ornamentals" that escaped to become invasive pests. Several of these are still being sold today.

One reason these once-desired species can displace native plants is because they **are** exotic. They were imported into this country without the natural complement of pests that keeps their populations in check at home (see chapter by Westbrooks and Eplee, this volume). Thus, one means of restoring balance to the invaded ecosystem is to seek biological control agents in the country of origin that can be imported to control the exotic pest. This is an expensive solution, because years of research are essential to ensure that the biological control agent

Fig. 2. Japanese honeysuckle strangles regenerating forests. Photo courtesy of Faith Campbell.

will not have unintended effects of its own. But where eradication is no longer possible, biological control may be a better long-term investment than continuing to fight these invaders with chemical weapons. In many situations, an integrated pest management program that uses both biological and pesticide weapons may be needed.

THINK GLOBALLY

Worldwide policies on plant movement have been developed to protect agricultural commodities from the threat of these exotic pests. Inspection services, such as those provided by the USDA Animal and Plant Health Inspection Service, and many quarantine regulations, are largely successful at preventing the introduction of agricultural crop pests. A list of known pests can be prepared for each crop, risks can be analyzed, and mitigating procedures can be prescribed for each pest deemed worthy of quarantine consideration (see chapter by Royer, this volume). Protecting complex native ecosystems presents a much more difficult problem because less information is available to help predict and detect new pests (see chapter by Wallner, this volume).

Regulating movement of flora and fauna is essential to prevent homogenization and subsequent irrevocable losses in biodiversity (see chapter by Westbrooks and White, this volume). International cooperation is vital to this process, but such cooperation can be achieved only by the honest application of scientific data. At present, science is often placed in a political balance with economic interests (see chapter by Campbell, this volume). The development and ratification of international standards, which would simplify some of the regulatory processes, is proceeding very slowly, in part because of an underlying conflict of philosophies among members of the World Trade Organization. Conservative members maintain that a commodity should be considered "guilty until proven innocent"; this means all commodities are prohibited until they are studied and granted a permit for importation. More liberal trading partners (including the US) have adopted an "innocent until proven guilty" approach, which some feel is too much like closing the barn door after the horse has escaped (2,7). One definition that must be agreed upon is "what makes a pest a quarantine pest?" The International Plant Protection Convention committee developing standards has recently proposed the following environmental impacts that might trigger quarantines:

1. reduction or elimination of endangered (or threatened) native plant species;

2. reduction or elimination of a keystone plant species (a species which plays a major role in the maintenance of an ecosystem);

3. reduction or elimination of a plant species which is a major component of a native ecosystem; causing a change to plant biological diversity in such a way to result in ecosystem destabilization; resulting in control, eradication or management programs that would be needed if a quarantine pest were introduced, and impacts of such programs (e.g., pesticides or release of non-indigenous predators and parasites) on biological diversity.

These drastic consequences seem reasonable triggers for quarantine, although one wonders how likely it is that we will have the data needed to prove a pest meets these criteria. But what about other more insidious pests? What about pests that are already widespread, but do not need a "boost" in their biodiversity? Should those pests also be regulated? Could we somehow stop, encourage, or cajole the nursery industry to stop selling exotic invasive weeds like Japanese honeysuckle, privet (Fig. 3), and Chinese wisteria? Why can't we test proposed imports of plants, reptiles, and amphibians for invasiveness, the way we test proposed biocontrol agents? Why can't we "close" a pathway known to leak pests without basing the quarantine on risk assessment data for one or more specific pests? Such a course is being developed for solid wood packing material (SWPM), because so many pests have recently caused great economic losses by moving in SWPM, e.g., the Asian longhorned beetle and the pinewood nematode. As explained in preceding chapters, at present, APHIS defines a quarantine pest as one either not present in the United States, or present but not widespread, and actively being controlled. To prevent imports, they must show a reasonable expectation that a commodity bears the pest, and estimate the risk to our resources if the pest were to become established. Then, they must develop the least costly, least restrictive control measure that will reduce the risk to a tolerable level, or be taken to court for trade protectionism at the WTO. Clearly, more international awareness, honest science, and discussion among top decisionmakers is needed to increase trust among nations so that we can protect our resources without fear of acrimony or reprisals. Cooperative research among nations concerned over particular pathways might help increase confidence in the results on both sides.

International cooperation at both the political and scientific levels is the obvious prerequisite to developing classical biological controls. This strategy requires extensive work in the pest's country of origin. A complex of regulating organisms often keeps a pest population in check in its native environment. The life cycles and host range of these regulating parasites and predators must be thoroughly studied in their country of origin. Only regulating organisms that are highly specific to the pest in question can be considered for importation as biocontrol agents. They must be screened for their ability to eat or infect all related crops. Only then can we safely permit importation to an accredited quarantine facility, where further screening is necessary to protect possible native hosts related to the target pest. Only after host specificity is ensured can permits be obtained for field releases.

It usually takes 5-10 years of research to find and adequately test potential biocontrol agents. Such programs can cost $5-$10 million, which seems expensive until one considers the alternatives: billions of dollars in lost resources and control costs, and irreplaceable losses in biodiversity (8).

Fig. 3. Chinese privet dominating floodplain understory at the State Botanical Gardens of Georgia.

An excellent example of strong interagency collaborations leading to successful biological control is the program initiated at Cornell University to control purple loosestrife. Purple loosestrife is an Eurasian wetland perennial plant introduced in the early 1800s, which is now choking many wetlands in temperate North America. In 1986, a multi-agency effort to develop biocontrol agents was initiated. After six years of testing, five insects were selected for release as biocontrol agents: a root-mining weevil, two leaf beetles, a flower-feeding weevil, and a seed-feeding weevil (1). This complex of insects is expected to reduce large populations of loosestrife, stabilize smaller populations, and reduce the probability of new outbreaks following disturbances. The program's goal is to reduce loosestrife abundance in North America to 10% of its current level, over 90% of the weed's present range.

Sometimes success comes slowly to biological control efforts. Tons of chemical insecticides have been sprayed over vast acreage to control the devastation of gypsy moth. Concern for human health, as well as the impact on non-target organisms, led to the development of biological control alternatives. A fungus named *Entomophaga maimaiga* was released in the early 1900s, with little apparent success. *Bacillus thuringiensis,* another biological agent, has been used extensively, but this agent also affects other moth and butterfly species. Later efforts centered on developing a viral agent, specific only to gypsy moth. The virus product can be used without harming other species of butterfly, but it is costly to produce because it must be propagated in live caterpillars. Scientists reintroduced *E. maimaiga* in 1985, but their efforts were frustrated because the gypsy moth population at the release location was decimated by the virus. Then, surprisingly in 1989, the fungus *E. maimaiga* became widely active, causing valuable reductions in gypsy moth populations. This outbreak, however, did not occur near the sites of reintroduction (9). If the active fungus is, in fact, the original isolate introduced and released nearly a century ago, did some unknown coincidence of necessary weather conditions permit it to spread and infect? Did it adapt to the U.S. environment? Or was an unknown and more virulent isolate of *E. maimaiga* accidentally introduced (14)?

Another slow success has begun after years of unflagging determination by researchers combating chestnut blight. Although early attempts to cross American chestnut with the blight-resistant Chinese chestnut were disappointingly slow, recent biotechnological advances have accelerated progress. Identification of genetic markers for the resistance genes now permit researchers to classify the offspring of their crosses as resistant or susceptible immediately, without waiting for the trees to mature. This breakthrough also reduces the chestnut planting that must be maintained for further crossing to a more affordable size (10). The American Chestnut Foundation now predicts that cultivars with blight resistance, and 15/16 American chestnut genes, will be available in the next 10 years.

In the 1950s, chestnut researchers in Italy found virus-like double-stranded RNA in the fungus, and the chestnut cankers began to heal. These infected "hypovirulent" strains have been much studied in the United States, but many of our fungal isolates refused to merge with the infected strains. Recently, researchers succeeded in getting synthetic DNA coding for the virus-like RNA particles into the spores of uninfected strains, so the particles may spread through sexual reproduction. The hope is that this hypovirulence will now spread naturally throughout the fungus population and lead to gradual debilitation of virulence (3,4).

After almost 10 years of negotiations, the United States and China have recently established a memorandum of understanding for cooperation in the area of biological control. Over 10,000 tiny wasps that parasitize *Oracella* have been sent to China and released in an attempt to control the mealybug (11). In exchange for this assistance, parasites of the hemlock wooly adelgid have been sent for study in the United States.

International cooperation must bridge many administrative and cultural differences that complicate effective communication, and the establishment of cooperative working relationships. But the rewards of such cooperation can extend even beyond the direct benefit to science and ecosystem health.

ACT LOCALLY

A close look at your own environment will likely reveal some exotic organism invading your favorite place. I predict that the more you look, the more you will see. I urge you to take action, at whatever level you can. Organize local efforts to increase awareness of the biological pollution threat. Support ongoing efforts by joining organizations such as The Nature Conservancy or Exotic Plant Pest Councils. Kill the exotic weeds in your own back yard. Choose a skirmish and begin to fight back, for the outcome of this war depends on each of us!!

GIVING AND GETTING HELP

I am convinced that, as they become familiar with the problems created by biological pollution, environmentalists will want to help in many ways. Of course, they can help by supporting the funding of organizations that are fighting biological pollution. But, funding increases are difficult to arrange in our times, and many people will want "hands on" experiences. I believe as awareness of this threat spreads, we are more likely to get people volunteering to help us in our work.

Help is a two-way street, however. It requires people who are willing to help and people who are willing to accept help. Too often research scientists fail to become personally involved in the application of their results on the ground. They feel they can get more science done by sticking to science. But researchers may find their resources dwindling as the federal pie shrinks. If we could accept doing less science, but persist in taking each major result all the way to a useful outcome, we would find eager hands waiting to help. Perhaps if the public saw and participated in the application of research results, our support base would grow rather than shrink. I challenge my colleagues, the many kinds of professionals dedicated to plant protection, to find ways to use willing volunteers in their work. We need help to deal with biological pollution, and we must learn to accept help.

Literature Cited

1.   Blossey, B. 1998. Biological control of purple loosestrife in North America. *In*: Exotic Pests of Eastern Forests. K.O. Britton, ed. Keen Impressions, Asheville, NC. 198 pp.

2.   Campbell, F.T. 2001. The science of risk assessment for phytosanitary regulation and the impact of changing trade regulations. BioScience 51:148-153.

3.   Chen, B., Choi, G.H., and Nuss, D.L. 1994. Attenuation of fungal virulence by synthetic infectious hypovirus transcripts. Science 264:1762-1764.

4.   Choi, G.H., and Nuss, D.L. 1992. Hypovirulence of chestnut blight fungus conferred by an infectious viral cDNA. Science 257:800-803.

5.   Kuhlman, E.G. 1978. The devastation of American chestnut by blight. Pp. 1-3 *in*: Proc. Am. Chestnut Symp. MacDonald, W.L., Check, F.C., Luchok, J., and Smith, H.C., eds. West Virginia Univ. Books, Morgantown. 122 pp.

6.   Liebhold, A.M., MacDonald, W.L., Bergdahl, D., and Mastro, V.C. 1995. Invasion by exotic forest pests: a threat to forest ecosystems. Forest Science Monograph 30. 49 pp.

7.   Mack, R.N., Simberloff, D., Lonsdale, W.M., Evans, H., Clout, M., and Bazzaz, F. 2000. Biotic invasions: causes, epidemiology, global consequences, and control. Issues in Ecology 5:1-25.

8.   Pimentel, D., Lach, L., Zuniga, R., and Morrison, D. 2000. Environmental and economic costs of non-indigenous species in the United States. BioScience 50:53-65.

9.   Reardon, R., and Hajek, A. 1993. *Entomophaga maimaiga* in North America: A Review. USDA Forest Service Northeastern Area For. Health Prot. NA-TP-15-93. 22 pp.

10.   Schlarbaum, S.E., Hebard, F., Spaine, P.C., and Kamalay, J.C. 1997. Three American tragedies: chestnut blight, butternut canker, and Dutch elm disease. *In*: Exotic Pests of Eastern Forests. Britton, K.O., ed. Keen Impressions, Asheville, NC. April 8-10, 1997. Pp. 45-54.

11.   Sun, J., DeBarr, G.L., Liu, T.X., Berisford, C.W., and Clarke, S.R. 1996. An unwelcome guest in China: a pine-feeding mealybug. J. of For. 94:27-32.

12.   Tasker, A.V. 1997. Length of lag-phase between introduction and spread of invasive plants. Weed Science Society of America, Regulatory Section, Abstracts of Annual Meeting 37:36.

13.   U.S. Congress, Office of Technology Assessment. Harmful Non-Indigenous Species in the United States. OTA-F-565 Washington, DC: U.S. Government Printing Office, September 1993.